5950

FUZZY LOGIC AND CONTROL

PRENTICE HALL SERIES ON ENVIRONMENTAL AND INTELLIGENT MANUFACTURING SYSTEMS
M. Jamshidi, Editor

Volume 1: *ROBOTICS AND REMOTE SYSTEMS FOR HAZARDOUS ENVIRONMENTS,* edited by M. Jamshidi and P. J. Eicker.

Volume 2: *FUZZY LOGIC AND CONTROL: Software and Hardware Applications,* edited by M. Jamshidi, N. Vadiee, and T. J. Ross.

Volume 3: *ARTIFICIAL INTELLIGENCE IN OPTIMAL DESIGN AND MANUFACTURING,* edited by Z. Dong.

FUZZY LOGIC AND CONTROL

Software and Hardware Applications

EDITORS
Mohammad Jamshidi
Nader Vadiee
Timothy J. Ross
University of New Mexico

PTR Prentice Hall, Englewood Cliffs, New Jersey 07632

Library of Congress Cataloging-in-Publication Date

Fuzzy logic and control: software and hardware applications/
  editors, MOHAMMAD JAMSHIDI, NADER VADIEE, TIMOTHY J. ROSS.
    p. cm. —(Environmental and intelligent manufacturing systems; v.2)
    Includes bibliographical references and indexes.
    ISBN 0-13-334251-4
    1. Intelligent control systems.   2. Fuzzzy systems.   3. Application software.
    I. Jamshidi, Mohammad. II. Vadiee, Nader   III. Ross, Timothy J.
    IV. Series
    TJ217.5.F89   1993   93-15427
    629.8'9   CIP

Editorial/production supervision: *Barbara Marttine*
Cover design: *Bruce Kenselaar*
Acquisitions editor: *Michael Hays*
Manufacturing buyer : *Mary McCartney*

©1993 by PTR Prentice-Hall, Inc.
A Simon & Schuster Company
Englewood Cliffs, New Jersey 07632

The publisher offers discounts on this book when ordered
in bulk quantities. For more information, contact::

Corporate Sales Department
PTR Prentice Hall
113 Sylvan Avenue
Englewood Cliffs, NJ 07632

Phone: 201-592-2863
Fax: 201-592-2249

All rights reserved. No part of this book may be
reproduced, in any form or by any means,
without permission in writing from the publisher.

Printed in the United States of America

10 9 8 7 6 5 4 3 2 1

ISBN 0-13-334251-4

Prentice-Hall International (UK) Limited, *London*
Prentice-Hall of Australia Pty. Limited, *Sydney*
Prentice-Hall Canada Inc., *Toronto*
Prentice-Hall Hispanoamericana, S.A., *Mexico*
Prentice-Hall of India Private Limited, *New Delhi*
Prentice-Hall of Japan, Inc., *Tokyo*
Simon & Schuster Asia Pte. Ltd., *Singapore*
Editora Prentice-Hall do Brasil, Ltda., *Rio de Janeiro*

THIS VOLUME IS DEDICATED

TO:

PROFESSOR LOTFI A. ZADEH
THE FATHER OF FUZZY LOGIC

WHO INSPIRED THE WORLD
WITH HIS CREATIVE WORKS ON

FUZZY SETS, FUZZY LOGIC,

APPROXIMATE REASONING,

AND FUZZY CONTROL.

TABLE OF CONTENTS

PREFACE..xi

1 Introduction..1
    M. Jamshidi

2 Set Theory – Fuzzy and Crisp Sets..10
    T. J. Ross

3 Propositional Calculus – Predicate Logic and Fuzzy Logic.......36
    T. J. Ross

4 Fuzzy Rule-Based Expert Systems - I..51
    N. Vadiee

5 Fuzzy Rule-Based Expert Systems - II...86
    N. Vadiee

6 Fuzzy Logic Software and Hardware..112
    M. Jamshidi

7 A Fuzzy Two-Axis Mirror Controller for
   Laser Beam Alignment..149
    R. D. Marchbanks

8 Introduction of Fuzzy Sets in Manufacturing Planning............171
    H. Fotouhie

9 A Comparison of Crisp and Fuzzy Logic Methods for
   Screening Enhanced Oil Recovery Techniques.........................181
    W. J. Parkinson and K. H. Duerre

10 A Fuzzy Logic Rule-Based System for Personnel Detection...217
    P. Sayka

11 Using Fuzzy Logic to Automatically Configure a
   Digital Filter..232
    R. J. Knight and M. T. Akbarzadeh

12 Simulation of Traffic Flow and Control Using
   Fuzzy and Conventional Methods..262
    R. L. Kelsey and K. R. Bisset

13	A Fuzzy Geometric Pattern Recognition Method with Learning Capability...279
	S. Peterson, C. Lashway, and D. Miller

14	Fuzzy Control of Robotic Manipulator.................................292
	K. K. Kumbla, J. Moya, and R. Baird

15	Use of Fuzzy Logic Control in Electrical Power Generation....329
	E. Kristjánsson, D. Barak, and K. Plummer

16	Fuzzy Logic Control of Resin Curing.......................................357
	M. McCullough

17	Fuzzy Logic Control in Flight Control Systems.......................363
	C. Baker

18	Tuning of Fuzzy Logic Controllers by Parameter Estimation Method...374
	N. K. Alang Rashid and A. S. Heger

SUBJECT INDEX..393

AUTHOR INDEX..395

* * * * *

FOREWORD

In the nearly four decades which have passed since the birth of Artificial Intelligence (AI), impressive progress has been made in our understanding of the basic issues relating to knowledge representation, automated reasoning, search techniques, game playing, and programming languages for AI. On closer analysis, however, a conclusion which emerges is that substantial progress was achieved in those areas in which classical symbolic logic—on which AI is based—provides an effective foundation for analysis and design. But what is widely acknowledged is that the traditional symbol-manipulation-oriented AI techniques have proved to be much less successful in the realms of commonsense reasoning, image understanding, speech recognition, handwriting recognition, summarization, learning from experience, and other fields in which pure symbol manipulation is a tool of very limited effectiveness.

Viewed in this perspective, the rapidly growing visibility of smart consumer products and systems—ranging from washing machines, air conditioners and camcorders to automobiles, subway trains, and mobile robots—reflects, in the main, the successful application of techniques derived from fuzzy logic (FL) and, more recently, neural network theory (NN) and probabilistic reasoning (PR), with the latter subsuming genetic algorithms (GA), belief networks, parts of learning theory, and chaotic systems. A distinguishing characteristic of such techniques—and especially that of FL—is that they aim at exploiting the tolerance for imprecision and uncertainty to achieve tractability, robustness, and low solution cost. In so doing, they mimic the remarkable ability of the human mind to learn and make rational decisions in an environment of uncertainty and imprecision.

It is convenient to view FL, NN, and PR as the principal constituents of what may be called soft computing (SC). In my view, it is only a matter of time before it will be widely recognized that AI should be based not on traditional, hard computing—which models modes of reasoning which are rigorous and precise—but on soft computing—a methodology of computing which attempts to come to grips with the pervasive imprecision and ill-definedness of the real world.

In the triumvirate of FL, NN, and PR, FL is concerned in the main with imprecision, NN with learning, and PR with uncertainty. In large measure, FL, NN, and PR are complementary rather than competitive. It is becoming increasingly clear that in many cases it is advantageous to employ FL, NN, and PR in combination rather than exclusively. A case in point is the growing number of neurofuzzy consumer products and systems which employ a combination of fuzzy logic and neural network techniques.

As one of the principal constituents of soft computing, fuzzy logic is playing a key role in the conception and design of what might be called high MIQ (Machine Intelligence Quotient) systems. There are two concepts within FL which play a central role in its applications. The first is that of a linguistic variable, that is, a variable whose values are words or sentences in a natural or synthetic language. The other is that of a fuzzy if-then rule in which the antecedent and consequent are propositions containing linguistic variables. The essential function served by linguistic variables is that of granulation of variables and their dependencies. In effect, the use of linguistic variables and fuzzy if-then rules results–through granulation–in soft data and compression which exploits the tolerance for imprecision and uncertainty. In this respect, fuzzy logic mimics the crucial ability of the human mind to summarize data and focus on decision-relevant information.

Although the basic ideas underlying fuzzy logic are simple and close to human intuition, there is a scarcity of books which present a readable introduction to fuzzy logic and describe its current applications. The volume produced by Professors Jamshidi and Ross and Dr. Vadiee deserves very high marks in both regards. All three editors have been in the forefront of instruction and research in fuzzy logic and have pioneered in the establishment of an active program in fuzzy logic and intelligent systems at the University of New Mexico. They and the contributors to this volume deserve our thanks and congratulations for producing a volume which provides both a lucid introduction to the theory and an authoritative and up-to-date description of many of its important applications.

Lotfi A. Zadeh

PREFACE

Intelligent control, which is defined as a combination of control theory, operations research, and artificial intelligence (AI) is emerging as one of the most popular new technologies in the industrial and manufacturing worlds. Among many possible new technologies based on AI, fuzzy logic is now perhaps the most popular area, judging by the billions of dollars worth of sales and close to 2000 patents issued in Japan alone since the announcement of the first fuzzy chips in 1987. Thanks to tremendous technological and commercial advances in fuzzy logic in Japan and other nations, today fuzzy logic is enjoying an unprecedented popularity in the technological and engineering fields including manufacturing. Fuzzy logic technology is now being used in numerous consumer and electronic products and systems, even in the stock market and medical diagnostics. The most important issue facing many industrialized nations in the next several decades will be global competition to an extent that has never before been posed. The arms race is diminishing and the economic race is in full swing. Fuzzy logic is but one such front for global technological, economical, and manufacturing competitions. An equally or perhaps much more important aspect of this new surge of interest in fuzzy logic is the educational aspect of fuzzy logic and fuzzy logic applications, including control systems. In 1989, in a study performed for the U.S. Congress, the United States Office of Technology Assessment studied more than 12 competing technologies for cost reduction in space applications. The number-one technology on their list turned out to be "expert systems," including "fuzzy expert systems."

The purpose of this book is to describe one experience in the education of engineering students at the University of New Mexico over a span of two years. First, the book provides some basic concepts of fuzzy set theory (Chapter 2), fuzzy logic (Chapter 3), fuzzy control (Chapters 3 and 4), and fuzzy logic software and hardware (Chapter 6). The softwares presented are Togai's Fuzzy-C Systems, NeuraLogix's NLX-230 Fuzzy Microcontroller, Bell Helicopter Textron's FULDEK, and University of New Mexico's FLCG. The reader may send in a postcard (found at the end of the book) to obtain further information on the latter two software packages. A number of actual software and hardware applications of fuzzy logic follow Chapter 6 as case studies in separate chapters (Chapters 7-18). The areas of application in the book are:

1. Laser beam alignment and tracking systems.
2. Manufacturing planning.
3. Enhanced oil recovery techniques.
4. Personnel detection.
5. Configuration of digital filters.
6. Traffic control.
7. Pattern recognition and genetic algorithms.
8. Robot manipulator control.
9. Flight control systems.
10. Electric power generation control.
11. Resin curing in industrial plants.

The reader may access the technical papers through an organized table of contents as well as the subject index at the end of the book. A list of all contributors is also available for easy access by readers of this volume.

The editors take this opportunity to thank Professor Lotfi Zadeh for his constant source of inspiration and encouragement. We thank all the authors for their contributions to fuzzy logic, to its applications, and to this volume. We also express our gratitude to Dean James Thompson, College of Engineering, Professor Nasir Ahmed, Chair, Electrical and Computer Engineering Department, and Professor Jerry Hall, Chair, Civil Engineering Department, all of the University of New Mexico, for supporting our efforts.

Our most sincere thanks and appreciation must go to Ms. Nancy Gillan of UNM's CAD Laboratory for Intelligent and Robotic Systems for her diligent and superb work in typesetting, copyediting, and preparing the manuscript for Prentice Hall Publishers. Last, but by no means least, we sincerely and thoughtfully thank our families for their understanding and sacrifice during the writing and assembling of this volume.

Mohammad Jamshidi
Nader Vadiee
Timothy J. Ross

CONTRIBUTORS

Akbarzadeh, Mohammad T.
 CAD Laboratory for Intelligent & Robotic Systems,
 University of New Mexico, Albuquerque, NM

Alang Rashid, Nahrul K.
 Chemical & Nuclear Engineering, University of New Mexico, Albuquerque, NM
 AND Nuclear Energy Unit, Malaysian Ministry of Energy, Kuala Lumpur, Malaysia

Baird, Ronald
 Electrical & Computer Engineering Department,
 University of New Mexico, Albuquerque, NM

Baker, Craig
 Honeywell Corporation, Defense Avionics Systems Division, Albuquerque, NM

Barak, Denis
 CAD Laboratory for Intelligent & Robotic Systems
 University of New Mexico, Albuquerque, NM

Bisset, Keith R.
 Los Alamos National Laboratory, Los Alamos, NM

Duerre, K. H.
 Los Alamos National Laboratory, Los Alamos, NM

Fotouhie, Hädie
 Mechanical Engineering Department, University of New Mexico, Albuquerque, NM

Heger, A. Sharif
 Chemical & Nuclear Engineering Department,
 University of New Mexico, Albuquerque, NM

Jamshidi, M.
 CAD Laboratory for Intelligent & Robotic Systems
 University of New Mexico, Albuquerque, NM

Kelsey, Robert L.
 Los Alamos National Laboratory, Los Alamos, NM

Knight, Robert J.
 Electrical & Computer Engineering Department,
 University of New Mexico, Albuquerque, NM

Kristjánsson, Erlendur
 CAD Laboratory for Intelligent & Robotic Systems
 University of New Mexico, Albuquerque, NM

Kumbla, Kishan Kumar
  CAD Laboratory for Intelligent & Robotic Systems
  University of New Mexico, Albuquerque, NM

Lashway, Clin
  CAD Laboratory for Intelligent & Robotic Systems
  University of New Mexico, Albuquerque, NM
  AND Raydec, Inc., Albuquerque, NM

Marchbanks, Richard D.
  Electrical & Computer Engineering Department
  University of New Mexico, Albuquerque, NM
  AND Los Alamos National Laboratory, Los Alamos, NM

McCullough, Mark
  Electrical & Computer Engineering Department
  University of New Mexico, Albuquerque, NM

Miller, Doug
  CAD Laboratory for Intelligent & Robotic Systems
  University of New Mexico, Albuquerque, NM

Moya, John
  Electrical & Computer Engineering Department
  University of New Mexico, Albuquerque, NM

Parkinson, W. J.
  Los Alamos National Laboratory, Los Alamos, NM

Peterson, Scott
  CAD Laboratory for Intelligent & Robotic Systems
  University of New Mexico, Albuquerque, NM
  AND Raydec, Inc., Albuquerque, NM

Plummer, Kenneth
  Electrical & Computer Engineering Department
  University of New Mexico, Albuquerque, NM

Ross, Timothy J.
  Civil Engineering Department, University of New Mexico, Albuquerque, NM

Sayka, Paul
  Los Alamos National Laboratory, Los Alamos, NM

Vadiee, Nader
  CAD Laboratory for Intelligent & Robotic Systems
  University of New Mexico, Albuquerque, NM

1
INTRODUCTION

Mo Jamshidi
University of New Mexico

One of the more popular new technologies is "intelligent control," which is defined as a combination of control theory, operations research, and artificial intelligence (AI). Among many possible new technologies based on AI, fuzzy logic is now perhaps the most popular area, judging by the billions of dollars worth of sales and close to 2000 patents issued in Japan alone since the announcement of the first fuzzy chips in 1987. Thanks to tremendous technological and commercial advances in fuzzy logic in Japan and other nations, today fuzzy logic is enjoying an unprecedented popularity in the technological and engineering fields including manufacturing. Fuzzy logic technology is now being used in numerous consumer and electronic products and systems, even in the stock market and medical diagnostics. The most important issue facing many industrialized nations in the next several decades will be global competition to an extent that has never before been posed. The arms race is diminishing and the economic race is in full swing. Fuzzy logic is but one such front for global technological, economical, and manufacturing competitions. An equally or perhaps much more important aspect of this new surge of interest in fuzzy logic is the educational aspect of fuzzy logic and fuzzy logic applications, including control systems. The purpose of this book is to describe one experience in the education of engineering students at the University of New Mexico (UNM) over a span of two years. The book will first provide some basic concepts of fuzzy set theory (Chapter 2), fuzzy logic (Chapter 3), fuzzy control (Chapters 3 and 4), and fuzzy logic software and hardware (Chapter 6). Then a number of actual software and hardware applications of fuzzy logic as case studies in separate chapters (Chapters 7-18). This chapter is an introduction to the entire book.

1.1　　BACKGROUND

In 1989, in a study performed for the U.S. Congress, the United States Office of Technology Assessment studied more than 12 competing technologies for cost reduction in space applications. The number-one technology on their list turned out to be "expert systems," including "fuzzy expert systems." Fuzzy logic and fuzzy expert control systems have found applications in numerous appliances and systems with moderate size and limited number of components. One of the first complex systems in which fuzzy logic has been successfully applied is cement kilns, which began in 1977 in Denmark. Today, most of the world's cement kilns are using some type of fuzzy expert system. The bulk of today's applications are undertaken by the Japanese industries. Ironically, fuzzy logic was first proposed by an American, Lotfi A. Zadeh, in 1965 when he published his seminal paper on "Fuzzy Sets." Zadeh showed that fuzzy logic is the foundation of any logic, regardless of how many truth values it may have. A fuzzy set has movable boundaries, i.e., the elements of such sets not only represent the colors black and white, but also allow a spectrum of gray colors in between.

Currently, one of the more active areas of fuzzy logic applications is control systems. Fuzzy controllers are expert control systems that smoothly interpolate between hard-boundary crisp rules. Rules fire simultaneously to continuous degrees or strengths and the multiple resultant actions are combined into an interpolated result. Processing of uncertain information and saving of energy using common sense rules and natural language statements are the basis for fuzzy control. The use of sensor data in practical control systems involves several tasks that are usually done by a human in the decision loop, e.g., an astronaut adjusting the position of a satellite or putting it in the proper orbit, a driver adjusting a vehicle's air-conditioning unit, etc. All such tasks must be performed based on the evaluation of data according to a set of rules in which the human expert has learned from experience or training. Often, if not most of the time, these rules are not crisp, i.e., some decisions are based on common sense or personal judgment. Such problems can be addressed by a set of fuzzy variables and rules which, if properly constructed, can make decisions as well as an expert.

This chapter represents the introduction to a book which has come out of the educational and research activities of a team from UNM over a period of two years. The structure of the chapter's presentation is as follows: Section 2 gives a brief introduction into fuzzy logic. An introduction to fuzzy control appears in Section 3. UNM's educational efforts since 1989 on fuzzy logic and fuzzy control are briefly described in Section 4, while Section 5 constitutes discussions on hardware and software for fuzzy logic. Some software (simulation) and hardware experiences (real-time experiments) are described in Section 6. The chapter will conclude with a section on the scope of this book.

1.2 FUZZY LOGIC

The need and use of multi-level logic can be traced from the ancient works of Aristotle, who is quoted as saying, "There will be a sea battle tomorrow." Such a statement is not yet true or false, but is potentially either. Much later, around AD 1285-1340, William of Occam supported two-values logic but speculated on what the true value of "if p then q" might be if one of the two components were neither true nor false. During the time period of 1878-1956, Lukasiewicz proposed a three-level logic as a "true" (1), a "false" (0), and a "neuter" (1/2), which represented half true or half false. In subsequent times, logicians in China and other parts of the world continued on the notion of multi-level logic and proposed multi-level logic. Zadeh, in his seminal 1965 paper [1], finished the task by following through the speculation of previous logicians and showing that what he called "fuzzy sets" were the foundation of *any* logic, regardless of the number of truth levels assumed. He chose the innocent word "fuzz" for the continuum of logical values between 0 (completely false) and 1 (completely true). The theory of fuzzy logic deals with two problems of: i) the fuzzy set theory, which deals with the ambiguity found in semantics, and ii) the fuzzy measurement theory, which deals with the ambiguous nature of judgments and evaluations.

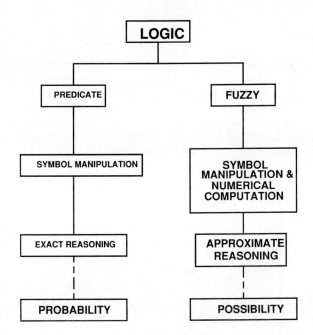

Figure 1-1. A comparison of predicate and fuzzy logic.

The primary motivation and "banner" of fuzzy logic is the possibility of exploiting tolerance for some inexactness and imprecision. Precision is often very costly, so if a problem does not require precision, one should not have to pay for it. The traditional example of parking a car is a noteworthy illustration. If the driver is not required to park the car within an exact distance from the curb, why spend any more time than necessary on the task as long as it is a legal parking operation. Fuzzy logic and classical logic differ in the sense that the former can handle both symbolic and numerical manipulation, while the latter can handle symbolic manipulation only (see Figure 1-1). In a broad sense, fuzzy logic is a union of fuzzy (fuzzified) crisp logics [2]. To quote Zadeh, "Fuzzy logic's primary aim is to provide a formal, computationally-oriented system of concepts and techniques for dealing with modes of reasoning which are approximate rather than exact." Thus, in fuzzy logic, exact (crisp) reasoning is considered to be the limiting case of approximate reasoning. In fuzzy logic one can see that everything is a matter of degree.

In an attempt to translate the crisp knowledge in a process, such as voltage across a terminal, to a linguistic or fuzzy knowledge -- that is, to go through the process of *"fuzzification"* -- one must make the binary input and output variables of a plant members of some fuzzy set, e.g. the set of bright images on the focal plane of a telescope or the set of small voltages across the armature of a DC motor. Fuzzy sets may be represented by a mathematical formulation often known as the membership function. This function gives a degree or grade of membership within the set. The membership function of a fuzzy set A , denoted by $\mu A(X)$, maps the elements of the universe X into a numerical value within the range [0,1], i.e.,

$$mA(X) : X \longrightarrow [0,1].$$

Note that a membership function is a so-called possibility function and not a probability function (see Chapter 2). Figure 1-2 shows some typical membership functions for various fuzzy (linguistic) variables (sets). Within this framework, a membership value of zero corresponds to a value which is definitely not an element of the fuzzy set, while a value of 1 corresponds to the case where the element is definitely a member of the set. In fuzzy logic, like binary logic, operations such as union, intersection, complement, OR, AND, etc., are all defined.

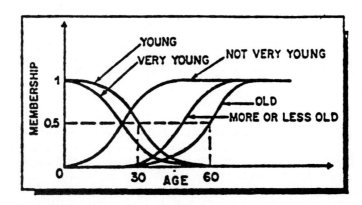

Figure 1-2. Typical membership functions for fuzzy sets or fuzzy variables.

1.3 FUZZY CONTROL

Fuzzy control systems are rule-based systems in which a set of so-called fuzzy rules represents a control decision mechanism to adjust the effects of certain causes coming from the system. The aim of fuzzy control systems is normally to substitute for or replace a skilled human operator with a fuzzy rule-based system. Figure 1-3 shows a block diagram for this definition. As shown, the human operator observes quantities by reading a meter or assessing a chart (i.e., noting fuzzy variables) and then performing a definite action, such as pushing a knob or turning a wheel (i.e., providing a crisp action or output y (see Figure 1-3). In a similar fashion the fuzzy controller, shown in Figure 1-3, used crisp data directly from a number of sensors; through the process of fuzzification these are changed to linguistic or fuzzy membership functions (fuzzified). They then go through a set of fuzzy "IF-THEN" rules in an inference engine much like an expert system and result in some fuzzy output(s) z. The fuzzy output(s) will then be changed back into crisp values through a process called *"defuzzification"* by some weighted average method such as the "centroidal" method (see Chapter 4). The result is a value of the output y*. In this way, the designer has obtained an approximate output y* for the actual output y.

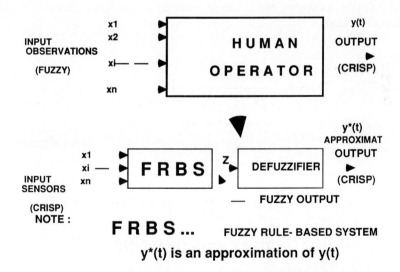

Figure 1-3. A definition of fuzzy control system

As indicated before, a fuzzy controller typically takes the form of a set of IF-THEN rules whose *antecedents* (IF part) and *consequents* (THEN part) are themselves membership functions. Consequents from different rules are numerically combined (typically union via MAX) and are then collapsed (typically taking the centroid of the combined distribution) to yield a single real-number (binary) output. Within the framework of a fuzzy expert system, like regular expert systems, typical rules can be the result of a human operator's knowledge, e.g.:

*"IF the Temperature is Hot
    THEN decrease the Current to a Medium level."*

In this rule, *Hot* and *Medium* are fuzzy variables. Such natural language rules can then be translated into typical computer language statements such as :

IF (A is A1 and B is B1 and C is C1 and D is D1)
    THEN (E is E1 and F is F1)

Using a set of rules such as these, an entire finite number of rules can be derived in the form of natural language statements as if a human operator were performing the controlling task. In any practical system, such as an air conditioning system, the user or an operator often fine-tunes, tweaks, and adjusts the knobs until the desired cool (or hot) air can be felt with the desired speed. Such operator knowledge can be utilized in the design of a fuzzy controller for an air conditioning unit system. One of the most common ways of designing a fuzzy controller is through "fuzzy rule-based systems." Figure 1-4 shows a typical fuzzy control architecture. The controller shows the processes of fuzzification, (i.e., binary to fuzzy transformation) and defuzzification (i.e., fuzzy to binary transformation).

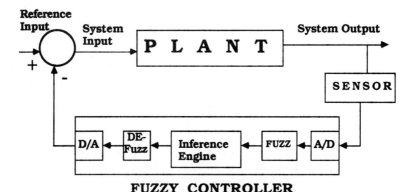

Figure 1-4. Block diagram for a typical fuzzy control system showing fuzzifier (FUZZ), defuzzifier (DE-Fuzz), and inference engine.

An alternative way of implementing a fuzzy control regime, which is similar to a conventional adaptive control law, is to use a standard crisp logic controller such as, say a PID (u = Kp.e(t) + K int e(t)dt + Kd de(t)/dt), and then use fuzzy IF-THEN rules to "tune" the gains of the conventional controller, i.e., Kp Ki, and Kd. Here variable e(t) is the tracking error for a system output y(t) which is to follow a desired output yd(t). Figure 1-5 shows one such adaptive fuzzy control architecture.

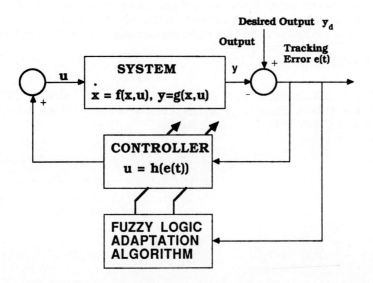

Figure 1-5. An adaptive (self-tuning) fuzzy control system.

The fuzzy control problem, like any control problem, is one of evaluating a mapping h(.), defined by:

$$h : e \longrightarrow u$$

where u and e are the control and the error signals, respectively. The choice of h(.) is the essence of the control problem. For a two-sensor case, i.e., a two-error variable (say e and Δe) and one control signal, the plot of u versus e and Δe will provide a surface this book will call *control surface*. Figure 1-6 shows two surfaces -- one (Figure 1-6[a]) belongs to an expert controller and the other (Figure 1-6[b]) represents a fuzzy controller. As seen here and indicated earlier, fuzzy controllers are expert control systems that smoothly interpolate between crisp rules. This often results in a savings of energy, because the fuzzy control surface fits underneath the expert control surface. Chapters 4 and 5 will treat fuzzy control in some detail.

1.4 SCOPE OF THE BOOK

The basic theme of this book is fuzzy logic with software and hardware applications. The entire book represents a result of teaching and research on fuzzy logic and its applications at UNM's CAD Laboratory for Intelligent and Robotic Systems for more than two years. The first six chapters represent an attempt to introduce readers to the basic and advanced concepts of fuzzy sets, fuzzy logic, fuzzy control, and fuzzy software and hardware. Chapter 6 provides a brief overview of available fuzzy logic software. These softwares are Togai's Fuzzy-C Systems, NeuraLogix's NLX-230 Fuzzy Microcontroller, Bell Helicopter Textron's FULDEK, and UNM's FLCG. At the end of the book is a postcard that readers may use to obtain further information on the latter two software packages. Chapters 7 through 18 present various software and hardware projects which either have been completed or were at a stage which initial results could be shared in open literature.

REFERENCES

1. Zadeh, L. A. (1965). "Fuzzy Sets." *Information and Control*, Vol. 8, pp. 335-353.
2. Zadeh, L. A. (1991). "Fuzzy Logic and the Calculus of Fuzzy If-Then Rules," *Proceedings of SYNAPSE '91*, Tokyo, Japan.

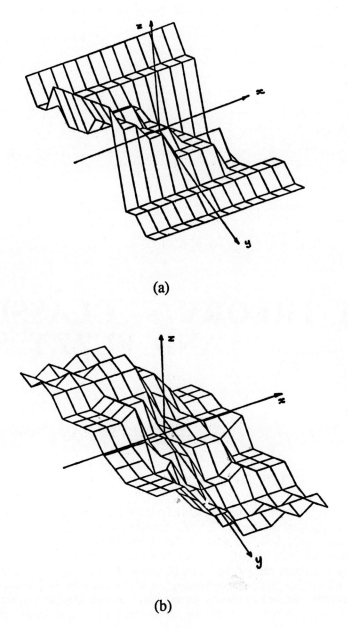

Figure 1-6. Two control surfaces: (a) an expert control surface, and (b) a fuzzy control surface.

2

SET THEORY — CLASSICAL AND FUZZY SETS

Timothy J. Ross
University of New Mexico

2.1   INTRODUCTION

Fuzzy set theory is developed here by comparing the precepts and operations of fuzzy sets with those of classical set theory. Fuzzy sets will be seen to contain the vast majority of the definitions, precepts, and axioms that define classical sets. In fact, very few differences exist between the two set theories. Fuzzy set theory is actually a fundamentally broader theory than current classical set theory, in that it considers an infinite number of "degrees of membership" in a set other than the canonical values of 0 and 1 apparent in classical set theory. In this sense, one could argue that classical sets are a limited form of fuzzy sets. Hence, it will be shown that fuzzy set theory is a comprehensive set theory.

Conceptually, a fuzzy set can be defined as a collection of elements in a universe of information where the boundary of the set contained in the universe is ambiguous, vague, and otherwise fuzzy. It is instructive to introduce fuzzy sets by first reviewing the elements of classical (crisp) set theory.

2.2 CLASSICAL SETS

Define X as the set of all objects with the same characteristics, called the universe of discourse, whose individual elements are denoted by x. The elements of the universe can be discrete and finite or continuous and infinite. Also, define the cardinality number, n_X, as the total number of elements in X. A set A consists of collections of some elements in X; set A is a subset of the universe X. Furthermore, the following notation holds:

$x \in X \rightarrow$ x belongs to X

$x \in A \rightarrow$ x belongs to A

$x \notin A \rightarrow$ x does not belong to A

For sets A and B on X, we also have

$A \subset B$ — A is contained in B
— $\forall x \in A$, then $x \in B$

$A \subseteq B$ — A is fully contained in B

$A = B$ — $A \subseteq B$ and $B \subseteq A$

We define the null set, $\emptyset$, as the set containing no elements, and the whole set, X, as the set of all elements in the universe. All subsets of X comprise a special set called the power set, P(X). For example, for the following universe X, the power set is enumerated.

Example: We have a universe comprised of three elements, X = {a, b, c}, so the cardinality is $n_X = 3$. The power set is:

P(X) = { $\emptyset$, {a}, {b}, {c}, {a, b}, {a, c}, {b, c}, {a, b, c} }

The cardinality of the power set, denoted $n_{P(X)}$, is found as:

$$n_{P(X)} = 2^{n_X} = 2^3 = 8$$

Note that if the cardinality of the universe is infinite, then the cardinality of the power set is also equal to infinity, i.e., If $n_X = \infty \rightarrow n_{P(X)} = \infty$.

Operations on Classical Sets

Let A and B be two sets on the universe X. The union between the two sets, denoted $A \cup B$, represents all those elements in the universe which reside in (or belong to) either the set A or the set B or both sets A and B. The intersection of the two sets, denoted $A \cap B$, represents all those elements in the universe X which reside in (or belong to) both sets A and B, simultaneously. The complement of a set A, denoted $\overline{A}$, is defined as the collection of all elements in the universe which do not reside in the set A. The difference of a set A with respect to B, denoted A\B, is defined as the collection of all elements in the universe which reside in A and which do not reside in B simultaneously. These operations are shown below.

Union: $A \cup B = \{x | x \in A \text{ or } x \in B\}$

Intersection: $A \cap B = \{x | x \in A \text{ and } x \in B\}$

Complement: $\overline{A} = \{x | x \notin A, x \in X\}$

Difference: $A \backslash B = \{x | x \in A \text{ and } x \notin B\}$

These four operations are shown below in terms of Venn diagrams in Figures 2-1 through 2-4.

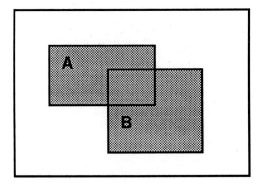

Figure 2-1. Union of sets A and B.

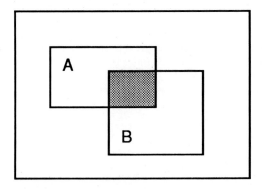

Figure 2-2. Intersection of sets A and B.

Chap. 2 Set Theory—Fuzzy and Crisp Sets

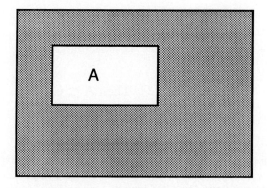

Figure 2-3. Complement of set A.

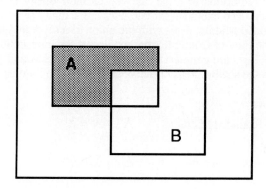

Figure 2-4. Difference operation A\B.

Properties of Classical (Crisp) Sets

Certain properties of sets are important to consider because of their influence on the mathematical manipulation of sets. The most appropriate properties for purposes of defining classical sets, to show their similarity to fuzzy sets, are:

Commutativity: $A \cup B = B \cup A$
$A \cap B = B \cap A$

Associativity: $A \cup (B \cup C) = (A \cup B) \cup C$
$A \cap (B \cap C) = (A \cap B) \cap C$

Distributivity: $A \cup (B \cap C) = (A \cup B) \cap (A \cup C)$
$A \cap (B \cup C) = (A \cap B) \cup (A \cap C)$

Idempotency: $A \cup A = A$
$A \cap A = A$

14   Chap. 2   Set Theory—Fuzzy and Crisp Sets

Identity:
$$A \cup \varnothing = A$$
$$A \cap X = A$$
$$A \cap \varnothing = \varnothing$$
$$A \cup X = X$$

Transitivity: If $A \subseteq B \subseteq C$, Then $A \subseteq C$

Involution: $\overline{\overline{A}} = A$

     Two special properties for set operations are known as the Excluded Middle Laws and DeMorgan's Laws. The excluded middle laws are very important properties because these are the only set operations described here that are not valid for both classical sets and fuzzy sets, while DeMorgan's laws are important because of their usefulness in proving tautologies and contradictions in logic, as well as their utility in a host of other set operations and proofs. The excluded middle laws are actually two laws: the first, called the Law of the Excluded Middle, deals with the union of a set A and its complement, and the second, called the Law of Contradiction, represents the intersection of a set A and its complement. These laws are enumerated below for two sets A and B, and DeMorgan's laws are displayed in the shaded areas of the Venn diagrams in Figures 2-5 and 2-6.

Excluded-Middle Laws:

     (1) Law of Excluded Middle $A \cup \overline{A} = X$

     (2) Law of Contradiction $A \cap \overline{A} = \varnothing$

DeMorgan's Laws:

$$\overline{(A \cap B)} = \overline{A} \cup \overline{B}$$

$$\overline{(A \cup B)} = \overline{A} \cap \overline{B}$$

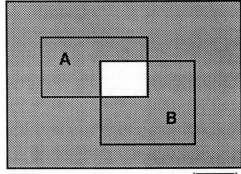

Figure 2-5. DeMorgan's Law $\overline{(A \cap B)}$.

Chap. 2   Set Theory—Fuzzy and Crisp Sets

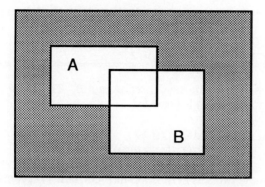

Figure 2-6. DeMorgan's Law $\overline{(A \cup B)}$.

Mapping of Classical Sets to Functions

Mapping is an important concept in relating set-theoretic forms to function-theoretic representations of information. Mapping in its most general form can be used to map elements or subsets on one universe of discourse to elements or sets in another universe. Suppose X and Y are two different universes of discourse (information). If an element x is contained in X and it corresponds to an element y contained in Y, this is generally termed a mapping from X to Y, or f: X→Y. As a mapping, the characteristic (indicator) function χ_A is defined by

$$\chi_A(x) = \begin{cases} 1, & x \in A \\ 0, & x \notin A \end{cases}$$

where χ_A expresses "membership" in set A for the element x in the universe. This membership idea is a mapping, from an element x in universe X to one of the two elements in universe Y; i.e., to the elements 0 or 1.

For any set A defined on the universe X, there exists a value-set, V(A) under the mapping of the characteristic function, χ. By convention the null set $\emptyset$ is assigned the membership value 0 and the whole set X is assigned the membership value 1.

Example: Continuing with the previous example of a universe with three elements, X = {a, b, c}, we desire to map the elements of the power set of X, i.e., P(X) to a universe, Y, consisting of only 2 elements (the characteristic function),

Y = {0, 1}

As before, the elements of the power set are enumerated below,

P(X) = { $\emptyset$, a, b, c, (a, b), (b, c), (a, c), (a, b, c)}

And now the elements in the value set V(A) as determined from the mapping are,

$$V[P(X)] = [\{0,0,0\}, \{1,0,0\}, \{0,1,0\}, \{0,0,1\}, \{1,1,0\},$$
$$\{0,1,1\}, \{1,0,1\}, \{1,1,1\}]$$

Now, define two sets on the universe X, sets A and B. The union of these two sets in terms of function-theoretic terms is given by,

Union: $A \cup B \rightarrow \chi_{A \cup B}(x) = \chi_A(x) \vee \chi_B(x)$
$= \max(\chi_A(x), \chi_B(x))$

The intersection of these two sets in function-theoretic terms is given by,

Intersection: $A \cap B \rightarrow \chi_{A \cap B}(x) = \chi_A(x) \wedge \chi_B(x)$
$= \min(\chi_A(x), \chi_B(x))$

The complement of a single set on universe X, say A, is given by,

Complement: $\overline{A} \ \text{\AE} \ c_{\overline{A}}(x) = 1 - c_A(x)$

For two sets on the same universe, say A and B, if one set (A) is contained in another set (B), then

Containment: $A \subseteq B \rightarrow \chi_A(x) \le \chi_B(x)$.

2.3 FUZZY SETS

In classical sets, or crisp sets, the transition between membership and non-membership in a given set for an element in the universe is abrupt and well-defined (said to be "crisp"). For an element in a universe which contains fuzzy sets this transition can be gradual. This transition among various degrees of membership can be thought of as conforming to the fact that the boundaries of the fuzzy sets are vague and ambiguous. Hence, membership of an element from the universe in this set is measured by a function which attempts to describe vagueness and ambiguity.

A fuzzy set then is a set containing elements which have varying degrees of membership in the set. This idea is contrasted with classical, or crisp, sets because members of a crisp set would not be members unless their membership was full or complete in that set (i.e., their membership is assigned a value of 1). Elements in a fuzzy set, because their membership can be a value other than complete, can also be members of other fuzzy sets on the same universe.

Elements of a fuzzy set are mapped to a universe of "membership values" using a function-theoretic form. Fuzzy sets are denoted by a set symbol with a tilde understrike; so for example, A̰ would be the "fuzzy set" A. This function maps elements of a fuzzy

Chap. 2   Set Theory—Fuzzy and Crisp Sets

set $\underset{\sim}{A}$ to a real numbered value on the interval 0 to 1. If an element in the universe, say x, is a member of fuzzy set $\underset{\sim}{A}$ then this mapping is given as,

$$\mu_{\underset{\sim}{A}}(x) \in [0,1]$$

$$\underset{\sim}{A} = (x, \mu_{\underset{\sim}{A}}(x) | x \in X)$$

These mappings are shown below in Figures 2-7 and 2-8 for crisp and fuzzy sets, respectively.

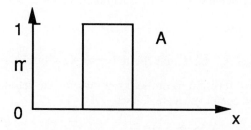

Figure 2-7. Membership function for crisp set A.

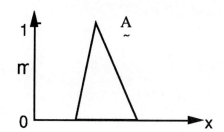

Figure 2-8. Membership function for fuzzy set $\underset{\sim}{A}$.

A notation convention for fuzzy sets that is popular in the literature when the universe of discourse, X, is discrete and finite, is given below for a fuzzy set $\underset{\sim}{A}$ by,

$$\underset{\sim}{A} = \frac{\mu_{\underset{\sim}{A}}(x_1)}{x_1} + \frac{\mu_{\underset{\sim}{A}}(x_2)}{x_2} + \cdots = \sum_i \frac{\mu_{\underset{\sim}{A}}(x_i)}{x_i}$$

and, when the universe, X, is continuous and infinite, the fuzzy set $\underset{\sim}{A}$ is denoted by,

$$A = \int \frac{\mu_A(x)}{x}$$

In both notations, the horizontal bar is not a quotient, but rather a delimiter. In both notations, the numerator in each individual expression is the membership value in set A associated with the element of the universe indicated in the denominator of each expression. In the first notation, the summation symbol is not for algebraic summation, but rather is denoting a fuzzy union; hence the "+" signs in the first notation are not algebraic "add," but rather function-theoretic union. In the second notation the integral sign is not an algebraic integral, but rather a set union notation for continuous variables.

Fuzzy Set Operations

Define three fuzzy sets A, B, C on the universe X. For a given element x of the universe, the following function theoretic operations for the set-theoretic operations of union, intersection, and complement are defined,

$$A, B, C \, on \, X$$

Union: $\mu_{A \cup B}(x) = \mu_A(x) \vee \mu_B(x)$

Intersection: $\mu_{A \cap B}(x) = \mu_A(x) \wedge \mu_B(x)$

Complement: $\mu_{\bar{A}}(x) = 1 - \mu_A(x)$

Venn diagrams for these operations, extended to consider fuzzy sets, are shown in Figures 2-9 through 2-11.

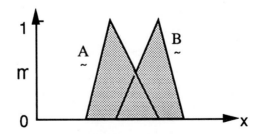

Figure 2-9. Union of fuzzy sets A and B.

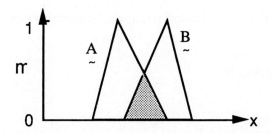

Figure 2-10. Intersection of fuzzy sets A and B.

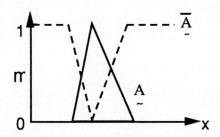

Figure 2-11. Complement of fuzzy set A.

Any fuzzy set A defined on a universe X is a subset of that universe. Also by definition, just as with classical sets, the membership value of any element x in the null set $\emptyset$ is 0 and the membership value of any element x in the whole set X is 1. Note that the null set and the whole set are not fuzzy sets in this context (no tilde understrike). The appropriate notation for these ideas is as follows:

$$A \subseteq X \rightarrow \mu_A(x) \leq \mu_x(x)$$

$$\forall x \in X, \mu_\phi(x) = 0$$

$$\forall x \in X, \mu_x(x) = 1$$

The collection of all fuzzy sets and fuzzy subsets on X is denoted as the fuzzy power set $P(X)$. It should be obvious to see, based on the fact that all fuzzy sets can overlap, that the cardinality of the fuzzy power set is infinite; that is,

$$\text{Cardinality of } P(x) \rightarrow n_{p(x)} = \infty$$

DeMorgan's laws for classical sets also hold for fuzzy sets, as denoted by the expressions below,

$$\overline{(A \cap B)} = \bar{A} \cup \bar{B}$$

$$\overline{(A \cup B)} = \bar{A} \cap \bar{B}$$

All other operations on classical sets, as enumerated before, also hold for fuzzy sets, except for the excluded middle laws. These two laws do not hold for fuzzy sets because of the fact that since fuzzy sets can overlap, a set and its complement can also overlap. The excluded middle laws, extended for fuzzy sets, are expressed by,

$$A \cup \bar{A} \neq X$$

$$A \cap \bar{A} \neq \phi$$

Extended Venn diagrams for these situations, and comparisons to the excluded middle laws for classical (crisp) sets, are shown below in Figures 2-12 and 2-13.

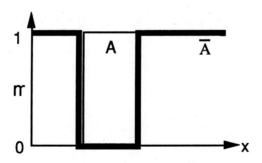

(a) Crisp set A and its complement.

Figure 2-12. Excluded Middle Laws for crisp sets.

Chap. 2 Set Theory—Fuzzy and Crisp Sets

Figure 2-12, continued:

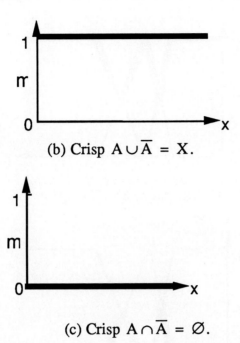

(b) Crisp $A \cup \overline{A} = X$.

(c) Crisp $A \cap \overline{A} = \emptyset$.

Figure 2-12. Excluded Middle Laws for crisp sets.

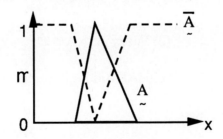

(a) Fuzzy set $\underset{\sim}{A}$ and its complement.

Figure 2-13. Excluded Middle Laws for fuzzy sets.

Figure 2-13, continued:

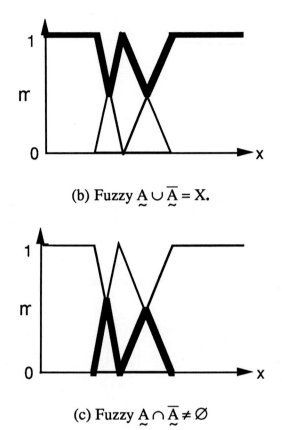

(b) Fuzzy $\underset{\sim}{A} \cup \overline{\underset{\sim}{A}} = X$.

(c) Fuzzy $\underset{\sim}{A} \cap \overline{\underset{\sim}{A}} \neq \emptyset$

Figure 2-13. Excluded Middle Laws for fuzzy sets.

Properties of Fuzzy Sets

Fuzzy sets follow the same properties as do crisp sets. Because of this and because the membership values of a crisp set are a subset of the interval [0,1], classical sets can be thought of as a special case of fuzzy sets. Frequently used properties of fuzzy sets are listed below.

Commutativity: $\underset{\sim}{A} \cup \underset{\sim}{B} = \underset{\sim}{B} \cup \underset{\sim}{A}$

$\underset{\sim}{A} \cap \underset{\sim}{B} = \underset{\sim}{B} \cap \underset{\sim}{A}$

Associativity:
$$A \cup (B \cup C) = (A \cup B) \cup C \text{ and } A \cap (B \cap C) = (A \cap B) \cap C$$

Distributivity:
$$A \cup (B \cap C) = (A \cup B) \cap (A \cup C) \text{ and } A \cap (B \cup C) = (A \cap B) \cup (A \cap B)$$

Idempotency: $A \cup A = A \text{ and } A \cap A = A$

Identity:
$$A \cup \emptyset = A \text{ and } A \cap X = A \text{ and } A \cap \emptyset = \emptyset \text{ and } A \cup X = X$$

Transitivity: If $A \subseteq B \subseteq C$, then $A \subseteq C$

Involution: $\overline{\overline{A}} = A$

Features of the Membership Function

Since all information contained in a fuzzy set is described by its membership function, it is useful to develop a lexicon of terms to describe various special features of this function. Figure 2-14 assists in this description.

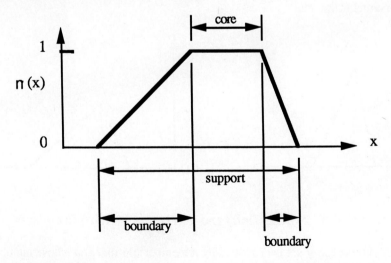

Figure 2-14. Core, support, and boundaries of a fuzzy set.

The core of a membership function for some fuzzy set $\underset{\sim}{A}$ is defined as that region of the universe that is characterized by complete and full membership in the set $\underset{\sim}{A}$. That is, the core is comprised of those elements of the universe, x, where $\mu_{\underset{\sim}{A}}(x) = 1$.

The support of a membership function for some fuzzy set $\underset{\sim}{A}$ is defined as that region of the universe that is characterized by non-zero membership in the set $\underset{\sim}{A}$. That is, the support is comprised of those elements of the universe, x, where $\mu_{\underset{\sim}{A}}(x) \neq 0$.

The boundaries of a membership function for some fuzzy set $\underset{\sim}{A}$ are defined as that region of the universe that contains elements that have a non-zero membership, but not complete membership. That is, the boundaries are comprised of those elements of the universe, x, where $0 < \mu_{\underset{\sim}{A}}(x) < 1$. These elements of the universe are those with some "degree" of fuzziness, or only partial membership in the fuzzy set $\underset{\sim}{A}$. Figure 2-14 illustrates the regions in the universe comprising the core, support, and boundaries of a typical fuzzy set.

A normal fuzzy set is one whose membership function has at least one element in the universe, x, whose membership value is unity. For fuzzy sets where one and only one element has a membership equal to one, this element is typically referred to as the "prototype" of the set, or the prototypical element. Figure 2-15 illustrates typical normal and non-normal fuzzy sets.

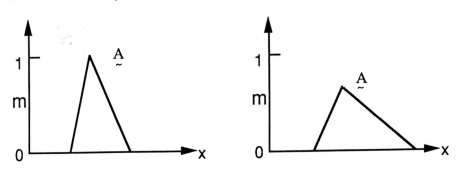

Figure 2-15. Normal (left) and non-normal (right) fuzzy sets.

A convex fuzzy set is described by a membership function whose membership values are strictly monotonically increasing, or whose membership values are strictly

Chap. 2 Set Theory—Fuzzy and Crisp Sets   25

monotonically decreasing, or whose membership values are strictly monotonically increasing then strictly monotonically decreasing with increasing values for elements in the universe. Said another way, if for all elements in a continuous fuzzy set $\underset{\sim}{A}$ where $x < y < z$, and where

$$\mu_{\underset{\sim}{A}}(y) \geq \min[\mu_{\underset{\sim}{A}}(x), \mu_{\underset{\sim}{A}}(z)]$$

then $\underset{\sim}{A}$ is said to be a convex fuzzy set. Figure 2-16 shows a typical convex fuzzy set and a typical non-convex fuzzy set.

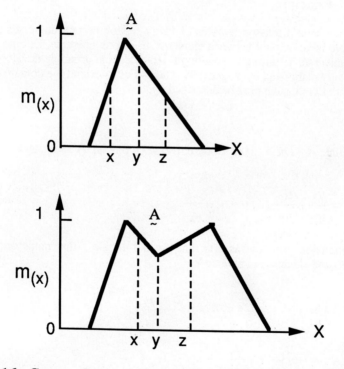

Figure 2-16. Convex fuzzy set (top) and non-convex fuzzy set (bottom).

It should be noted, that a special property of two convex fuzzy sets, say $\underset{\sim}{A}$ and $\underset{\sim}{B}$, is that the intersection of these two convex fuzzy sets is also a convex fuzzy set, as shown in Figure 2-17. That is, for $\underset{\sim}{A}$ and $\underset{\sim}{B}$, which are both convex, $\underset{\sim}{A} \cap \underset{\sim}{B}$ is also convex.

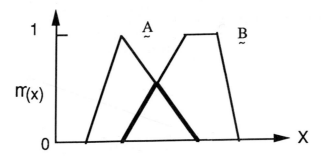

Figure 2-17. The intersection of two convex fuzzy sets.

Extension Principle

Heretofore we have discussed features of fuzzy sets on certain universes of discourse. Suppose there is a mapping between elements, u, of one universe, U, onto elements, v, of another universe, V, through a function f. An extension principle developed by Zadeh (1975) and later elaborated by Yager (1986) allows us to extend the domain of a function on fuzzy sets. Let this mapping be described by,

Let $f: u \to v$

Define $\underset{\sim}{A}$ to be a fuzzy set on universe U; that is, $\underset{\sim}{A} \subset U$, and

$$\underset{\sim}{A} = \frac{\mu_1}{u_1} + \frac{\mu_2}{u_2} + \cdots\cdots\cdots \frac{\mu_n}{u_n}$$

Then the extension principle asserts that, for a function f that maps one element in universe U to one element in universe V,

$$f\left(\underset{\sim}{A}\right) = f\left(\frac{\mu_1}{u_1} + \frac{\mu_2}{u_2} + \cdots\cdots\cdots + \frac{\mu_n}{u_n}\right)$$

$$= \frac{\mu_1}{f(u_1)} + \frac{\mu_2}{f(u_2)} + \cdots\cdots\cdots + \frac{\mu_n}{f(u_n)}$$

This mapping is said to be "one-to-one."

Example: Let a fuzzy set $\underset{\sim}{A}$ be defined on the universe $U = \{1, 2, 3\}$. We wish to map elements of this fuzzy set to another universe, V, under the function

$v = f(u) = 2u-1.$

Chap. 2 Set Theory—Fuzzy and Crisp Sets

We see that the elements of V are, V={1, 3, 5}. So, for example suppose the fuzzy set $\underset{\sim}{A}$ is given by,

$$\underset{\sim}{A} = \frac{0.6}{1} + \frac{1}{2} + \frac{0.8}{3},$$

then the fuzzy membership function for v = f(u), = 2u-1 would be,

$$f\left(\underset{\sim}{A}\right) = \frac{0.6}{1} + \frac{1}{3} + \frac{0.8}{5}$$

For cases where this functional mapping f maps products of elements from two universes, say U_1 and U_2, to another universe V, and we define $\underset{\sim}{A}$ as a fuzzy set on the Cartesian space $U_1 \times U_2$, then

$$f\left(\underset{\sim}{A}\right) = \left\{ \frac{\min[\mu_1(i), \mu_2(j)]}{f(i,j)} \setminus i \in U_1, j \in U_2 \right\}$$

where $\mu_1(i)$ and $\mu_2(j)$ are the separable membership projections of $\mu(i,j)$ from the Cartesian space $U_1 \times U_2$ when $\mu(i,j)$ can not be determined. This projection involves the invocation of a condition known as "non-interaction" between the separate universes, and is analogous to the assumption of independence employed in probability theory which reduces a joint probability density function to the product of its separate marginal density functions. In the fuzzy non-interaction case we use the minimum function as opposed to the product function used in probability theory.

Example: Suppose we have the integers 1 to 10 as the elements of two identical but different universes; let

$U_1 = U_2 = \{1, 2, 3, \ldots, 10\}$, then define two fuzzy sets $\underset{\sim}{A}$ and $\underset{\sim}{B}$ on universe U_1 and U_2, respectively,

Define $\underset{\sim}{A} = \underset{\sim}{2} = "approximately\ 2" = \frac{0.6}{1} + \frac{1}{2} + \frac{0.8}{3}$ and

define $\underset{\sim}{B} = \underset{\sim}{6} = "approximately\ 6" = \frac{0.8}{5} + \frac{1}{6} + \frac{0.7}{7}$,

then the product of ("approximately 2") x ("approximately 6") should map to a fuzzy number "approximately 12," which is a fuzzy set defined on a universe, say V, of integers, V={5, 6, ..., 18, 21}, as determined by the extension principle,

$$2\underset{\sim}{x}6 = \left(\frac{0.6}{1} + \frac{1}{2} + \frac{0.8}{3}\right) x \left(\frac{0.8}{5} + \frac{1}{6} + \frac{0.7}{7}\right)$$

$$= \frac{\min(0.6, 0.8)}{5} + \frac{\min(0.6, 1)}{6} + \ldots\ldots + \frac{\min(0.8, 1)}{18} + \frac{\min(0.8, 0.7)}{21}$$

$$= \frac{0.6}{5} + \frac{0.6}{6} + \frac{0.6}{7} + \frac{0.8}{10} + \frac{1}{12} + \frac{0.7}{14} + \frac{0.8}{15} + \frac{0.8}{18} + \frac{0.7}{21}$$

The complexity of the extension principle increases when we consider if more than the combination of the input variables, U_1 and U_2, are mapped to the same variable in the output space, V. In this case we take the maximum membership grades of the combinations mapping to the same output variable, or for the mapping shown below we get,

$$\mu_{\underset{\sim}{A}}(u_1, u_2) = \max\left[\min\{\mu_1(u_1), \mu_2(u_2)\}\right]$$

$$v = f(u_1, u_2)$$

Example: We have two fuzzy sets $\underset{\sim}{a}$ and $\underset{\sim}{b}$, each defined on its own universe, as follows,

$$\underset{\sim}{a} = \frac{0.2}{1} + \frac{1}{2} + \frac{0.7}{4} \text{ and } \underset{\sim}{b} = \frac{0.5}{1} + \frac{1}{2}$$

We wish to determine the membership values for the mapping

$$f(\underset{\sim}{a}, \underset{\sim}{b}) = \underset{\sim}{a} x \underset{\sim}{b}$$

$$= \frac{\min(0.2, 0.5)}{1} + \frac{\max[\min(0.2, 1), \min(0.5, 1)]}{2} + \frac{\max[\min(0.7, 0.5), \min(1, 1)]}{4} + \frac{\min(0.7, 1)}{8}$$

$$= \frac{0.2}{1} + \frac{0.5}{2} + \frac{1}{4} + \frac{0.7}{8}$$

In this case, the mapping involves two ways to produce a 2 (1x2 and 2x1) and two ways to produce a 4 (4x1 and 2x2).

2.4   RELATIONS AMONG CLASSICAL (CRISP) SETS

The Cartesian product of two universes X and Y is determined as

Chap. 2 Set Theory—Fuzzy and Crisp Sets

$$XxY = \{(x,y) / x \in X, y \in Y\}$$

which combines $\forall x \in X \text{ and } \forall y \in Y$ in an ordered pair and forms <u>unconstrained</u> matches between x and y. That is, every element in universe X is related completely to every element in universe Y. The "strength" of this relationship between ordered pairs of elements in each universe is measured by the characteristic function, where a value of unity is associated with "complete relationship" and a value of zero is associated with "no relationship," i.e., the binary values 1 and 0. An example is given in the Sagittal diagram shown below, and in the matrix expression to follow, where values of unity in the relation matrix, denoted R, correspond to the ordered pairs of mappings in the relation.

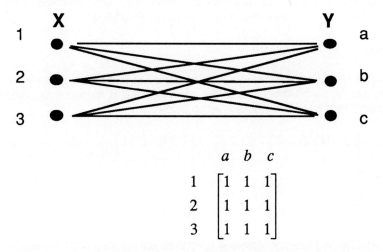

Figure 2-18. Sagittal diagrams (top) and matrix expressions (bottom) of an unconstrained relation.

A more general crisp relation, R, exists when matches between elements in two universes are constrained. Again, the characteristic function is used to assign values of relationship in the mapping of the Cartesian space X x Y to the binary values of (0,1).

$$\chi_R(x,y) = \begin{cases} 1, & (x,y) \in R \\ 0, & (x,y) \notin R \end{cases}$$

Example: In many biological models, members of certain species can reproduce only with certain members of another species. Hence, only some elements in two or more universes are paired. An example is shown below for two 2-member species.

$$R = \{(1,a),(2,b)\} \qquad R \subset X \times Y \qquad R = \begin{matrix} \\ 1 \\ 2 \end{matrix} \begin{matrix} a & b \\ \begin{bmatrix} 1 & 0 \\ 0 & 1 \end{bmatrix} \end{matrix}$$

Cardinality of Crisp Relations

Suppose n elements of the universe X are related (paired) to m elements of the universe Y. If the cardinality of X is n_X and the cardinality of Y is n_Y, then the cardinality of the relation, R, between these two universes is $n_{X*Y} = n_X * n_Y$. The cardinality of the power set describing this relation, $P(X*Y)$ is then $n_{P(X*Y)} = 2^{(n_X * n_Y)}$.

Operations on Crisp Relations

Define R and S as two separate relations on the Cartesian universe X x Y, and define the null relation and the complete relation as the matrices O and E, respectively. An example of a 4x4 form of the O and E matrices is given below.

$$O = \begin{bmatrix} 0 & 0 & 0 & 0 \\ 0 & 0 & 0 & 0 \\ 0 & 0 & 0 & 0 \\ 0 & 0 & 0 & 0 \end{bmatrix} \qquad E = \begin{bmatrix} 1 & 1 & 1 & 1 \\ 1 & 1 & 1 & 1 \\ 1 & 1 & 1 & 1 \\ 1 & 1 & 1 & 1 \end{bmatrix}$$

The following function-theoretic operations for the two crisp relations (R, S) can now be defined.

Union: $\quad R \cup S \rightarrow \chi_{R \cup S}(x,y) = \max[\chi_R(x,y), \chi_S(x,y)]$

Intersection: $\quad R \cap S \rightarrow \chi_{R \cap S}(x,y) = \min[\chi_R(x,y), \chi_S(x,y)]$

Complement: $\quad \overline{R} \rightarrow \chi_{\overline{R}}(x,y) = 1 - \chi_R(x,y)$

Containment: $\quad R \subset S \rightarrow \chi_R(x,y) \leq \chi_S(x,y)$

Identity: $\quad (\emptyset \rightarrow O \text{ and } X \rightarrow E)$

Further, the operations of commutativity, associativity, distributivity, involution, and idempotency all hold just as they do for classical set operations. Moreover, DeMorgan's laws and the Excluded Middle laws also hold for crisp (classical)

Chap. 2   Set Theory—Fuzzy and Crisp Sets   31

relations just as they do for crisp (classical) sets. The null relation, O, and the complete relation, E, are analogous to the null set and the whole set in set-theoretic form.

Now let R be a relation that relates, or maps, elements from universe X to universe Y, and let S be a relation that relates, or maps, elements from universe Y to universe Z. Can we find a relation, T, that relates the same elements in universe X that R handles to the same elements in universe Z that S handles? The answer is yes, and we do this with an operation known as <u>composition</u>. So for the Sagittal diagram in Figure 2-19, we wish to find a relation T that relates the ordered pair (X_1, Z_2); i.e., $(X_1, Z_2) \in T$.

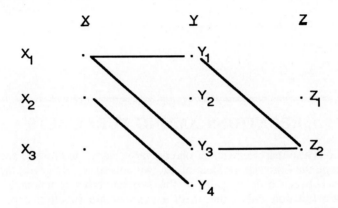

Figure 2-19. Sagittal diagram relating elements of three universes.

There are two common forms of the composition operation; one is called the max-min composition and the other is termed the max-product composition. The max-min composition is defined by the expression

$$T = R \circ S$$
$$\chi_T(x,z) = \bigvee_{y \in Y} (\chi_R(x,y) \wedge \chi_S(y,z))$$

and the max-product (sometimes called max-dot) composition is defined by the expression,

$$T = R \circ S$$
$$\chi_T(x,z) = \bigvee_{y \in Y} (\chi_R(x,y) \cdot \chi_S(y,z))$$

Example: The matrix expression for the crisp relations shown in Figure 2-19 can be found using the max-min composition operation. Relations R and S would be expressed as,

$$R = \begin{array}{c} \\ x_1 \\ x_2 \\ x_3 \end{array} \begin{array}{cccc} y_1 & y_2 & y_3 & y_4 \\ \begin{bmatrix} 1 & 0 & 1 & 0 \\ 0 & 0 & 0 & 1 \\ 0 & 0 & 0 & 0 \end{bmatrix} \end{array} \quad \text{and} \quad S = \begin{array}{c} \\ y_1 \\ y_2 \\ y_3 \\ y_4 \end{array} \begin{array}{cc} z_1 & z_2 \\ \begin{bmatrix} 0 & 1 \\ 0 & 0 \\ 0 & 1 \\ 0 & 0 \end{bmatrix} \end{array}$$

The resulting relation T would then be determined by max-min composition as,

$$T = \begin{array}{c} \\ x_1 \\ x_2 \\ x_3 \end{array} \begin{array}{cc} z_1 & z_2 \\ \begin{bmatrix} 0 & 1 \\ 0 & 0 \\ 0 & 0 \end{bmatrix} \end{array}$$

2-5 RELATIONS AMONG FUZZY SETS

Fuzzy relations also map elements of one universe, say X, to those of another universe, say Y, through the Cartesian product of the two universes. However, the "strength" of the relation between ordered pairs of the two universes is not measured with the characteristic function, but rather with a membership function expressing various "degrees" of the strength of the relation on the unit interval [0,1]. Hence, a fuzzy relation $\underset{\sim}{R}$, is a mapping from the Cartesian space X x Y to the interval [0,1], where the strength of the mapping is expressed by the membership function of the relation for ordered pairs from the two universes, or $\mu_{\underset{\sim}{R}}(x,y)$.

Cardinality of Fuzzy Relations

Since the cardinality of fuzzy sets on any universe is infinity, then the cardinality of a fuzzy relation between two or more universes is also infinity.

Operations on Fuzzy Relations

Let $\underset{\sim}{R}$, $\underset{\sim}{S}$, and $\underset{\sim}{T}$ be fuzzy relations on the Cartesian space X x Y. Then the following operations apply,

Union: $\qquad \mu_{\underset{\sim}{R} \cup \underset{\sim}{S}}(x,y) = \max\left(\mu_{\underset{\sim}{R}}(x,y), \mu_{\underset{\sim}{S}}(x,y)\right)$

Intersection: $\mu_{R\cap S}(x,y) = \min\left(\mu_R(x,y), \mu_S(x,y)\right)$

Complement: $\mu_{\bar{R}}(x,y) = 1 - \mu_R(x,y)$

Containment: $R \subset S \Rightarrow \mu_R(x,y) \leq \mu_S(x,y)$

Just as for crisp relations, the operations of commutativity, associativity, distributivity, involution, and idempotency all hold for fuzzy relations. Moreover, DeMorgan's laws hold for fuzzy relations just as they do for crisp (classical) relations, and the null relation, O, and the complete relation, E, are analogous to the null set and the whole set in set-theoretic form, respectively. The operations that do not hold for fuzzy relations, as is the case for fuzzy sets in general, are the Excluded Middle laws. Since a fuzzy relation $\underset{\sim}{R}$ is also a fuzzy set, there is overlap between a relation and its complement, and

$$\underset{\sim}{R} \cup \overline{\underset{\sim}{R}} \neq E$$

$$\underset{\sim}{R} \cap \overline{\underset{\sim}{R}} \neq O$$

As seen, the Excluded Middle laws for relations do not result in the null relation, O, or the complete relation, E.

Because fuzzy relations in general are fuzzy sets, we can define the Cartesian product between fuzzy sets. Let $\underset{\sim}{A}$ be a fuzzy set on universe X and $\underset{\sim}{B}$ be a fuzzy set on universe Y; then the Cartesian product between fuzzy sets $\underset{\sim}{A}$ and $\underset{\sim}{B}$ will result in a fuzzy relation $\underset{\sim}{R}$, or

$$\underset{\sim}{A} \times \underset{\sim}{B} = \underset{\sim}{R} \subset X \times Y$$

with membership function,

$$\mu_R(x,y) = \mu_{A \times B}(x,y) = \min\left(\mu_A(x), \mu_B(y)\right)$$

Example: Suppose we have two fuzzy sets on a universe, $\underset{\sim}{A}$ and $\underset{\sim}{B}$, and we want to find the fuzzy Cartesian product between them. Let,

$$\underset{\sim}{A} = \frac{0.2}{x_1} + \frac{0.5}{x_2} + \frac{1}{x_3} \quad \text{and} \quad \underset{\sim}{B} = \frac{0.3}{y_1} + \frac{0.9}{y_2}$$

Then the fuzzy Cartesian product is,

$$\underset{\sim}{A} \times \underset{\sim}{B} = \underset{\sim}{R} = \begin{array}{c} \\ x_1 \\ x_2 \\ x_3 \end{array} \begin{array}{c} z_1 \quad z_2 \\ \begin{bmatrix} 0.2 & 0.2 \\ 0.3 & 0.5 \\ 0.3 & 0.9 \end{bmatrix} \end{array}$$

Fuzzy composition can be defined just as it is for crisp (binary) relations. Suppose $\underset{\sim}{R}$ is a fuzzy relation on the Cartesian space X x Y, $\underset{\sim}{S}$ is a fuzzy relation on Y x Z, and $\underset{\sim}{T}$ is a fuzzy relation on X x Z; then the fuzzy max-min composition is defined as:

Let $\underset{\sim}{T} = \underset{\sim}{R} \circ \underset{\sim}{S}$

$$\mu_{\underset{\sim}{T}}(x,z) = \bigvee_{y \in Y} \left(\mu_{\underset{\sim}{R}}(x,y) \wedge \mu_{\underset{\sim}{S}}(y,z) \right)$$

and the fuzzy max-product composition is defined as,

$$\mu_{\underset{\sim}{T}}(x,z) = \bigvee_{y \in Y} \left(\mu_{\underset{\sim}{R}}(x,y) \bullet \mu_{\underset{\sim}{S}}(y,z) \right)$$

Example: Let us extend the information contained in the Sagittal diagram shown in Figure 2-19 to include fuzzy relationships between the universes X-Y (denoted by the fuzzy relation $\underset{\sim}{R}$) and Y-Z (denoted by the fuzzy relation $\underset{\sim}{S}$). Consider the following fuzzy relations,

$$R = \begin{bmatrix} 0.7 & 0.5 \\ 0.8 & 0.4 \end{bmatrix} \text{ and } S = \begin{bmatrix} 0.9 & 0.6 & 0.2 \\ 0.1 & 0.7 & 0.5 \end{bmatrix}$$

Then the resulting relation, T, which relates elements of universe X to elements of universe Z, can be found by max-min composition to be,

$$T = \begin{bmatrix} 0.7 & 0.6 & 0.5 \\ 0.8 & 0.6 & 0.4 \end{bmatrix}$$

It should be pointed out that neither crisp nor fuzzy compositions have inverses in general; that is

$$R \circ S \neq S \circ R$$

This result is general for any matrix operation, fuzzy or otherwise, which must satisfy consistency between the cardinal counts of elements in respective universes. Even for the case of square matrices, the composition inverse is not guaranteed.

REFERENCES

Yager, R. R. (1986) "A characterization of the extension principle," *Fuzzy Sets and Systems*, **18**, 205-217.

Zadeh, L. A. (1975) "The concept of a linguistic variable and its application to approximate reasoning," *Information Sciences*, **8**, 199-249; and also **9**, 43-80.

3

PROPOSITIONAL CALCULUS — PREDICATE LOGIC AND FUZZY LOGIC

Timothy J. Ross
University of New Mexico

3.1 PREDICATE LOGIC

In classical predicate logic, a simple proposition, P, is a linguistic statement contained within a universe of propositions which can be identified as being strictly true or strictly false. The veracity (truth) of the proposition, P, can be assigned a binary truth value, called T(P), just as an element in a universe is assigned a binary quantity to measure its membership in a particular set. For binary (Boolean) predicate logic, T(P) is assigned a value of 1 (truth) or 0 (false). If U is the universe of all propositions, then T is a mapping of these propositions to the binary quantities (0, 1), or

$$T: U \to \{0,1\}$$

Now let P and Q be two simple propositions on the same universe of discourse that can be combined using the following five logical connectives,

Chap. 3   Predicate Logic and Fuzzy Logic   37

(i)   disjunction   ($\vee$)
(ii)   conjunction   ($\wedge$)
(iii) negation   ($-$)
(iv)   implication   ($\rightarrow$)
(v)   equality   ($\leftrightarrow$ or $\equiv$)

to form logical expressions involving the two simple propositions. These connectives can be used to form new propositions from simple propositions.

Now define sets A and B from universe X, where these sets might represent linguistic ideas or thoughts. Then a *propositional calculus* will exist for the case where proposition P measures the truth of the statement that an element, x, from the universe X is contained in set A and the truth of the statement that this element, x, is contained in set B, or more conventionally
   P: truth that $x \in A$
   Q: truth that $x \in B$,   where truth is measured in terms of the truth value, i.e.,

   If $x \in A$, T(P) = 1; otherwise T(P) = 0.
   If $x \in B$, T(Q) = 1; otherwise T(Q) = 0, or using the characteristic function to represent truth (1) and false (0),

$$\chi_A(x) = \begin{cases} 1, x \in A \\ 0, x \notin A \end{cases}$$

Example: Let P be the proposition "Carol is a woman" and let Q be the proposition "Carol is pregnant." Let X be the universe of people, x is an element (Carol), A is the set of all women, and B is the set of all pregnant women. Hence,
   P: x is in A
   Q: x is in B.

The five logical connectives defined above can be used to create compound propositions, where a compound proposition is defined as a logical proposition formed by logically connecting two or more simple propositions. Just as we are interested in the truth of a simple proposition, predicate logic also involves the assessment of the truth of compound propositions. For the case of two simple propositions, the resulting compound propositions are defined below in terms of their binary truth values,

$P: x \in A$, $\overline{P}: x \notin A$

$P \vee Q \Rightarrow x \in A \text{ or } B$

   Hence, $T(P \vee Q) = \max(T(P), T(Q))$

$P \wedge Q \Rightarrow x \in A \text{ and } B$

   Hence, $T(P \wedge Q) = \min(T(P), T(Q))$

$If\ T(P) = 1,\ then\ T(\overline{P}) = 0;\ If\ T(P) = 0,\ then\ T(\overline{P}) = 1$

$P \leftrightarrow Q \Rightarrow x \in A, B$

$$Hence,\ T(P \leftrightarrow Q) = T(P)$$
$$= T(Q)$$

The logical connective "implication" presented here is also known as the classical implication, to distinguish it from an alternative form due to Lukasiewicz, a Polish mathematician in the 1930s, who was first credited with exploring logics other than Aristotelian (classical or binary logic) logic. This classical form of the implication operation requires some explanation.

For a proposition P defined on set A and a proposition Q defined on set B, the implication "P implies Q" is equivalent to taking the union of elements in the complement of set A with the elements in the set B. That is, the logical implication is analogous to the set-theoretic form,

$P \rightarrow Q \equiv \overline{A} \cup B\ is\ true \equiv either\ "not\ in\ A"\ or\ "in\ B"$

So that $(P \rightarrow Q) \leftrightarrow (\overline{P} \vee Q)$

$$T(P \rightarrow Q) = T(\overline{P} \vee Q) = \max(T(\overline{P}), T(Q))$$

This is linguistically equivalent to the statement, "P implies Q is true" when either "not A" or "B" is true. Graphically this implication and the analogous set operation is represented by the Venn diagram in Figure 3-1. As noted in the diagram, the region represented by the difference A\B is the set region where the implication "P implies Q" is false (the implication "fails"). The shaded region in Figure 3-1 represents the collection of elements in the universe where the implication is true.

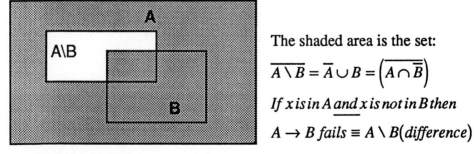

The shaded area is the set:

$\overline{A \setminus B} = \overline{A} \cup B = \overline{(A \cap \overline{B})}$

$If\ x\ is\ in\ A\ and\ x\ is\ not\ in\ B\ then$

$A \rightarrow B\ fails \equiv A \setminus B(difference)$

Figure 3-1. Graphical analog of the classical implication operation.

Chap. 3 Predicate Logic and Fuzzy Logic

Now, with two propositions (P and Q) each being able to take on one of two truth values (true or false, 1 or 0), there will be a total of $2^2 = 4$ propositional situations. These situations are illustrated, along with the appropriate truth values, for the propositions P and Q and the various logical connectives between them in the truth table below.

P	Q	$\bar{P}$	$P \vee Q$	$P \wedge Q$	$P \rightarrow Q$	$P \leftrightarrow Q$
T(1)	T(1)	F(0)	T(1)	T(1)	T(1)	T(1)
T(1)	F(0)	F(0)	T(1)	F(0)	F(0)	F(0)
F(0)	T(1)	T(1)	T(1)	F(0)	T(1)	F(0)
F(0)	F(0)	T(1)	F(0)	F(0)	T(1)	T(1)

Suppose the implication operation involves two different universes of discourse; P is a proposition described by set A, which is defined on universe X, and Q is a proposition described by set B, which is defined on universe Y. Then the implication "P implies Q" can be represented in set-theoretic terms by the relation R, where R is defined by

$$R = (A \times B) \cup (\bar{A} \times Y) \equiv \text{IF A, THEN B}$$

If $x \in A$ where $x \in X, A \subset X$

then $y \in B$ where $y \in Y, B \subset Y$

This implication is also equivalent to the linguistic rule form: IF A, THEN B. The graphic shown below in Figure 3-2 represents the Cartesian space of the product X x Y, showing typical sets A and B, and superposed on this space is the set-theoretic equivalent of the implication. That is,

$$P \rightarrow Q \Rightarrow \text{If } x \in A, \text{ then } x \in B, \text{ or } P \rightarrow Q \equiv \bar{A} \cup B$$

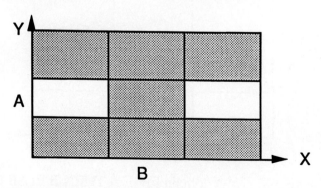

Figure 3-2. The Cartesian space showing the implication IF A, THEN B.

The shaded regions of the compound Venn diagram in Figure 3-2 represent the truth domain of the implication, IF A, THEN B (P implies Q).

Another compound proposition in linguistic rule form is the expression,

IF A, THEN B, ELSE C.

In predicate logic this has the form,

$$(P \rightarrow Q) \vee (\overline{P} \rightarrow S)$$

where $P: x \in A, A \subset X$
$Q: y \in B, B \subset Y$
$S: y \in C, C \subset Y$

Linguistically, this compound proposition could be expressed as,

IF A, THEN B, or
IF $\overline{A}$, THEN C.

The set-theoretic equivalent of this compound proposition is given by,

IF A THEN B ELSE C $\equiv (AxB) \cup (\overline{A}xC) = R = $ relation on $X \times Y$

The graphic shown in Figure 3-3 illustrates the shaded region representing the truth domain for this compound proposition.

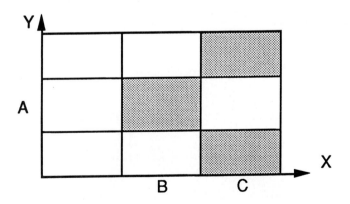

Figure 3-3. Truth domain for IF A THEN B ELSE C.

Chap. 3   Predicate Logic and Fuzzy Logic

Tautologies

In predicate logic it is useful to consider compound propositions that are always true, irrespective of the truth values of the individual simple propositions. Classical logical compound propositions with this property are called "tautologies." Tautologies are useful for deductive reasoning and for making deductive inferences. So, if a compound proposition can be expressed in the form of a tautology, the truth value of that compound proposition is known to be true. Inference schemes in expert systems often employ tautologies. The reason for this is that tautologies are logical formulas that are true on logical grounds alone.

One of these, known as Modus Ponens deduction, is a very common inference scheme used in forward chaining rule-based expert systems. It is an operation whose task is to find the truth value of a consequent in a production rule, given the truth value of the antecedent in the rule. Modus Ponens deduction concludes that, given two propositions, a and a-implies-b, both of which are true, then the truth of the simple proposition b is automatically inferred. Another useful tautology is the Modus Tollens inference, which is used in backward-chaining expert systems. In Modus Tollens an implication between two propositions is combined with a second proposition and both are used to imply a third proposition. Some common tautologies are listed below.

$$\overline{B} \cup B \leftrightarrow X \qquad (A \wedge (A \rightarrow B)) \rightarrow B \quad \text{(Modus Ponens)}$$

$$A \cup X; \quad \overline{A} \cup X \leftrightarrow X \qquad (\overline{B} \wedge (A \rightarrow B)) \rightarrow \overline{A} \,\text{(Modus Tollens)}$$

$$A \leftrightarrow B$$

A proof of the truth value of the Modus Ponens deduction is listed here.

Proof:
$$(A \wedge (A \rightarrow B)) \rightarrow B$$
$$(A \wedge (\overline{A} \cup B)) \rightarrow B$$
$$((A \wedge \overline{A}) \cup (A \wedge B)) \rightarrow B$$
$$(\phi \cup (A \wedge B)) \rightarrow B$$
$$(A \wedge B) \rightarrow B$$
$$\overline{(A \wedge B)} \cup B$$
$$(\overline{A} \vee \overline{B}) \cup B$$

$$\overline{A} \vee (\overline{B} \cup B)$$
$$\overline{A} \cup X$$

$$X \to T(X) = 1; \quad \underline{QED}$$

A similar display of the truth value of this tautology is shown below in truth table form.

A	B	A→B	(A∧(A→B))	(A∧(A→B))→B
0	0	1	0	1
0	1	1	0	1
1	0	0	0	1
1	1	1	1	1

A proof of the truth value of the Modus Tollens inference is listed here.

Proof:
$$(\overline{B} \wedge (A \to B)) \to \overline{A}$$
$$(\overline{B} \wedge (\overline{A} \cup B)) \to \overline{A}$$
$$((\overline{B} \wedge \overline{A}) \cup (\overline{B} \wedge B)) \to \overline{A}$$
$$((\overline{B} \wedge \overline{A}) \cup \phi) \to \overline{A}$$
$$(\overline{B} \wedge \overline{A}) \to \overline{A}$$
$$\overline{(\overline{B} \wedge \overline{A})} \cup \overline{A}$$
$$(\overline{\overline{B}} \vee \overline{\overline{A}}) \cup \overline{A}$$
$$B \cup (A \cup \overline{A})$$
$$B \cup X = X \to T(X) = 1$$

Chap. 3 Predicate Logic and Fuzzy Logic

A similar display of the truth value of this tautology is shown below in truth table form.

A	B	$\overline{A}$	$\overline{B}$	A→B	$(\overline{B} \wedge (A \rightarrow B))$	$(\overline{B} \wedge (A \rightarrow B)) \rightarrow \overline{A}$
0	0	1	1	1	1	1
0	1	1	0	1	0	1
1	0	0	1	0	0	1
1	1	0	0	1	0	1

Contradictions

Compound propositions that are always false, regardless of the truth value of the individual simple propositions comprising the compound proposition, are called contradictions. Some simple contradictions are listed below.

$$\overline{B} \cap B$$
$$A \cap \phi ; \overline{A} \cap \phi$$

Deductive Inferences

The Modus Ponens deduction is used as a tool for inferencing in rule-based systems. A typical IF-THEN rule is use to determine whether an antecedent (cause or action) infers a consequent (effect or reaction). Suppose we have a rule of the form,

> IF A, THEN B

This rule could be translated into a relation using the Cartesian product of sets A and B, that is:

> R = A x B.

Now suppose a new antecedent, say A' is known. Can we use Modus Ponens deduction to infer a new consequent, say B', resulting from the new antecedent? That is, in rule form

> IF A', THEN B' ?

The answer, of course, is yes, through the use of the composition relation. Since "A implies B" is defined on the Cartesian space X x Y, B' can be found through the following set-theoretic formulation,

$$B' = A' \circ R = A' \circ \left((A \times B) \cup (\overline{A} \times Y)\right)$$

Modus Ponens deduction can also be used for the compound rule,

IF A, THEN B, ELSE C

using the relation defined as,

$$R = (A \times B) \cup (\overline{A} \times C)$$

For this compound rule, if we define another antecedent A', the following possibilities exist, depending on: (i) whether A' is fully contained in the original antecedent A, (ii) whether A' is contained only in the complement of A, or (iii) whether A' and A overlap to some extent as shown below,

If $A' \subset A$, then $y = B$

If $A' \subset \overline{A}$, then $y = C$

If $A' \cap A \neq \phi, A' \cap \overline{A} \neq \phi, y = B \cup C$

To define the implication operation or a tautology in function-theoretic terms we need to define the truth value of a universe of discourse. For a universe Y, we define

T(Y) = 1 and T(ϕ) = 0.

The rule, IF A, THEN B (P defined on set A in universe X, and Q defined on set B in universe Y) is then defined in function-theoretic terms as,

$$P \rightarrow Q \Rightarrow R = (A \times B) \cup (\overline{A} \times Y)$$

$$\chi_R(x,y) = \max\left[(\chi_A(x) \wedge \chi_B(y)), ((1 - \chi_A(x)) \wedge 1)\right], \text{ where } \chi(\cdot) \text{ is}$$

the characteristic function as defined before.

Example: Suppose we have two universes of discourse described by the following collection of elements, X = {1, 2, 3, 4} and Y = {1, 2, 3, 4, 5, 6}. Define crisp set A on universe X and crisp set B on universe Y as follows, A = {2, 3} and B = {3, 4}. The deductive inference IF A, THEN B then yields the following matrix describing $\chi_R(x,y)$.

Chap. 3   Predicate Logic and Fuzzy Logic

$$R = \begin{array}{c} \\ 1 \\ 2 \\ 3 \\ 4 \end{array} \begin{array}{c} 1\ 2\ 3\ 4\ 5\ 6 \\ \begin{bmatrix} 1 & 1 & 1 & 1 & 1 & 1 \\ 0 & 0 & 1 & 1 & 0 & 0 \\ 0 & 0 & 1 & 1 & 0 & 0 \\ 1 & 1 & 1 & 1 & 1 & 1 \end{bmatrix} \end{array}$$

The compound rule IF A, THEN B, ELSE C is defined as

$$R = (A \times B) \cup (A \times C) \Rightarrow (P \rightarrow Q) \vee (\overline{P} \rightarrow S) \text{ and we have}$$

$$\chi_R(x,y) = \max\left[(\chi_A(x) \wedge \chi_B(y)),((1-\chi_A(x)) \wedge \chi_C(y))\right]$$

Example: Continuing with the previous example, suppose we define a crisp set C on universe Y as C = {6, 7}. The deductive inference IF A THEN B ELSE C then yields the following matrix describing $\chi_R(x,y)$.

$$R = \begin{array}{c} \\ 1 \\ 2 \\ 3 \\ 4 \end{array} \begin{array}{c} 1\ 2\ 3\ 4\ 5\ 6 \\ \begin{bmatrix} 0 & 0 & 0 & 0 & 1 & 1 \\ 0 & 0 & 1 & 1 & 0 & 0 \\ 0 & 0 & 1 & 1 & 0 & 0 \\ 0 & 0 & 0 & 0 & 1 & 1 \end{bmatrix} \end{array}$$

3.2   FUZZY LOGIC

A fuzzy logic proposition, $\underset{\sim}{P}$, is a statement involving some concept without clearly defined boundaries. Linguistic statements that tend to express subjective ideas and which can be interpreted slightly differently by various individuals typically involve fuzzy propositions. Most natural language is fuzzy, in that it involves vague and imprecise terms. Statements describing a person's height or weight, or assessments of people's preferences about colors or menus can be used as examples of fuzzy propositions. The truth value assigned to $\underset{\sim}{P}$ can be any value on the interval [0,1]. The assignment of the truth value to a proposition is actually a mapping from the interval [0,1] to the universe U of truth values, T,

$$T: U \rightarrow [0,1]$$

As in classical binary logic, we assign a logical proposition to a set in the universe of discourse. Fuzzy propositions are assigned to fuzzy sets. Suppose proposition P is assigned to fuzzy set $\underset{\sim}{A}$, then the truth value of a proposition, denoted $T(\underset{\sim}{P})$, is given by

$$T\left(\underset{\sim}{P}\right) = \mu_{\underset{\sim}{A}}(x) \text{ where } 0 \leq \mu_{\underset{\sim}{A}} \leq 1$$

The degree of truth for the proposition $P: x \in \underset{\sim}{A}$ is equal to the membership grade of x in $\underset{\sim}{A}$.

The logical connectives of negation, disjunction, conjunction, and implication are also defined for a fuzzy logic. These connectives are given below for two simple propositions: P defined on fuzzy set $\underset{\sim}{A}$ and Q defined on fuzzy set $\underset{\sim}{B}$.

Negation:

$$T(\overline{\underset{\sim}{P}}) = 1 - T(\underset{\sim}{P})$$

Disjunction:
$$\underset{\sim}{P} \vee \underset{\sim}{Q} \Rightarrow x \text{ is } \underset{\sim}{A} \text{ or } \underset{\sim}{B}$$

$$T(\underset{\sim}{P} \vee \underset{\sim}{Q}) = \max\left(T(\underset{\sim}{P}), T(\underset{\sim}{Q})\right)$$

Conjunction:
$$\underset{\sim}{P} \wedge \underset{\sim}{Q} \Rightarrow x \text{ is } \underset{\sim}{A} \text{ and } \underset{\sim}{B}$$

$$T(\underset{\sim}{P} \wedge \underset{\sim}{Q}) = \min\left(T(\underset{\sim}{P}), T(\underset{\sim}{Q})\right)$$

Implication:
$$\underset{\sim}{P} \rightarrow \underset{\sim}{Q} \Rightarrow x \text{ is } \underset{\sim}{A}, \text{ then } x \text{ is } \underset{\sim}{B}$$

$$T(\underset{\sim}{P} \rightarrow \underset{\sim}{Q}) = T(\overline{\underset{\sim}{P}} \vee \underset{\sim}{Q}) = \max\left(T(\overline{\underset{\sim}{P}}), T(\underset{\sim}{Q})\right)$$

As before in binary logic, the implication connective can be modeled in rule based form,

$$\underset{\sim}{P} \rightarrow \underset{\sim}{Q} \quad \text{is: IF } x \text{ is } \underset{\sim}{A}, \text{ THEN } y \text{ is } \underset{\sim}{B}$$

Chap. 3   Predicate Logic and Fuzzy Logic

and it is equivalent to the following fuzzy relation, $\underset{\sim}{R}$,

$$\underset{\sim}{R} = \left(\underset{\sim}{A} \times \underset{\sim}{B}\right) \cup \left(\overline{\underset{\sim}{A}} \times Y\right)$$

whose membership function is expressed by the following formula,

$$\mu_{\underset{\sim}{R}}(x,y) = \max\left[\left(\mu_{\underset{\sim}{A}}(x) \wedge \mu_{\underset{\sim}{B}}(y)\right), \left(1 - \mu_{\underset{\sim}{A}}(x)\right)\right]$$

Example: Suppose we are judging an Olympic gymnastics event and we are using two metrics to make our decisions regarding scoring the performance of each gymnast. Our metrics are the "form" of the gymnast, denoted universe X = {1, 2, 3, 4}, and the "degree of difficulty" of the gymnast's maneuvers, denoted universe Y = {1, 2, 3, 4, 5, 6}. In both universes the lowest numbers are the "best form" and the "highest difficulty." A contestant by the name of Gazelda has just received scores of "medium form," denoted by fuzzy set $\underset{\sim}{A}$, and "medium difficulty," denoted fuzzy set $\underset{\sim}{B}$. We wish to determine the implication of such a result, i.e., IF $\underset{\sim}{A}$, THEN $\underset{\sim}{B}$. The judges assign Gazelda the following fuzzy sets to represent her scores:

$$\underset{\sim}{A} = \text{medium form} = \frac{0.6}{2} + \frac{1}{3} + \frac{0.2}{4}$$

$$\underset{\sim}{B} = \text{medium difficulty} = \frac{0.4}{2} + \frac{1}{3} + \frac{0.8}{4} + \frac{0.3}{5}$$

The following matrices are then determined in developing the membership function of the implication, $\mu_{\underset{\sim}{R}}(x,y)$.

$$\underset{\sim}{A} \times \underset{\sim}{B} = \begin{array}{c} \\ 1 \\ 2 \\ 3 \\ 4 \end{array} \begin{array}{cccccc} 1 & 2 & 3 & 4 & 5 & 6 \\ \left[\begin{array}{cccccc} 0 & 0 & 0 & 0 & 0 & 0 \\ 0 & 0.4 & 0.6 & 0.6 & 0.3 & 0 \\ 0 & 0.4 & 1 & 0.8 & 0.3 & 0 \\ 0 & 0.2 & 0.2 & 0.2 & 0.2 & 0 \end{array}\right] \end{array}$$

$$\overline{A} \times Y = \begin{array}{c} \\ 1 \\ 2 \\ 3 \\ 4 \end{array} \begin{array}{cccccc} 1 & 2 & 3 & 4 & 5 & 6 \\ \left[\begin{array}{cccccc} 1 & 1 & 1 & 1 & 1 & 1 \\ 0.4 & 0.4 & 0.4 & 0.4 & 0.4 & 0.4 \\ 0 & 0 & 0 & 0 & 0 & 0 \\ 0.8 & 0.8 & 0.8 & 0.8 & 0.8 & 0.8 \end{array} \right] \end{array}$$

and finally, $R = \max(A \times B, \overline{A} \times Y)$

$$R = \begin{array}{c} \\ 1 \\ 2 \\ 3 \\ 4 \end{array} \begin{array}{cccccc} 1 & 2 & 3 & 4 & 5 & 6 \\ \left[\begin{array}{cccccc} 1 & 1 & 1 & 1 & 1 & 1 \\ 0.4 & 0.4 & 0.6 & 0.6 & 0.4 & 0.4 \\ 0 & 0.4 & 1 & 0.8 & 0.3 & 0 \\ 0.8 & 0.8 & 0.8 & 0.8 & 0.8 & 0.8 \end{array} \right] \end{array}$$

When the logical conditional implication is of the compound form,

IF x is A, THEN y is B, ELSE y is C

then the equivalent fuzzy relation, R, is expressed as,

$$R = \left(A \times B \right) \cup \left(\overline{A} \times C \right)$$

whose membership function is expressed by the following formula,

$$\mu_R(x,y) = \max\left[\left(\mu_A(x) \wedge \mu_B(y) \right), \left(1 - \mu_A(x) \wedge \mu_C(y) \right) \right]$$

3.3 APPROXIMATE REASONING

The ultimate goal of fuzzy logic is to form the theoretical foundation for reasoning about imprecise propositions; such reasoning is referred to as approximate reasoning. Approximate reasoning is analogous to predicate logic for reasoning with precise propositions, and hence is an extension of classical propositional calculus that deals with partial truths.

Chap. 3 Predicate Logic and Fuzzy Logic

Suppose we have a rule-based format to represent fuzzy information. These rules are expressed in conventional antecedent-consequent form, such as,

<u>Rule-1</u>: IF x is $\underset{\sim}{A}$, THEN y is $\underset{\sim}{B}$, where $\underset{\sim}{A}$ and $\underset{\sim}{B}$ represent fuzzy propositions (sets)

Now suppose we introduce a new antecedent, say $\underset{\sim}{A'}$, and we consider the following rule,

<u>Rule-2</u>: IF x is $\underset{\sim}{A'}$, THEN y is $\underset{\sim}{B'}$

From information derived from Rule-1, is it possible to derive the consequent in Rule-2, $\underset{\sim}{B'}$? The answer is yes, and the procedure is fuzzy composition. The consequent $\underset{\sim}{B'}$ can be found from the composition operation,

$$\underset{\sim}{B'} = \underset{\sim}{A'} \circ \underset{\sim}{R}$$

Example: Continuing with the gymnastics example, suppose that the fuzzy relation just developed, i.e., $\underset{\sim}{R}$, describes Gazelda's abilities, and we wish to know what "degree of difficulty" would be associated with a form score of: "almost good form." That is, with a new antecedent, $\underset{\sim}{A'}$, the following consequent, $\underset{\sim}{B'}$, can be determined using composition. Let,

$$\underset{\sim}{A'} = \text{almost good form} = \frac{0.5}{1} + \frac{1}{2} + \frac{0.3}{3}$$

then $\underset{\sim}{B'} = \underset{\sim}{A'} \circ \underset{\sim}{R} = \begin{bmatrix} 0.5 & 0.5 & 0.6 & 0.6 & 0.5 & 0.5 \end{bmatrix}$

or, alternatively, $\underset{\sim}{B'} = \frac{0.5}{1} + \frac{0.5}{2} + \frac{0.6}{3} + \frac{0.6}{4} + \frac{0.5}{5} + \frac{0.5}{6}$

In other words, the consequent is fairly diffuse, where there is no strong (weak) membership value for any of the difficulty scores (i.e., no membership values near 0 or 1).

An interesting issue in approximate reasoning is the idea of an inverse relationship between fuzzy antecedents and fuzzy consequences arising from the composition operation. Consider the following problem. Suppose we use the original antecedent, $\underset{\sim}{A}$, in the fuzzy composition. Do we get the original fuzzy consequent, $\underset{\sim}{B}$, as a result of the operation? That is, does

$$\underset{\sim}{B} = \underset{\sim}{A} \circ \underset{\sim}{R}?$$

The answer is an unqualified no, and one should not expect an inverse to exist for fuzzy composition.

Example: Again, continuing with the gymnastics example, suppose that $\underset{\sim}{A'} = \underset{\sim}{A} =$ "medium form," then

$$\underset{\sim}{B'} = \underset{\sim}{A'} \circ \underset{\sim}{R} = \underset{\sim}{A} \circ \underset{\sim}{R} = \frac{0.4}{1} + \frac{0.4}{2} + \frac{1}{3} + \frac{0.8}{4} + \frac{0.4}{5} + \frac{0.4}{6} \neq \underset{\sim}{B}$$

That is, the new consequent does not yield the original consequent ($\underset{\sim}{B}$=medium difficulty) because the inverse is not guaranteed with fuzzy sets.

In classical binary logic this inverse does exist, that is, crisp Modus Ponens would give:

$$B' = A' \circ R = A \circ R = B$$

where the sets A and B are crisp, and the relation R is also crisp.

In the case of approximate reasoning, the fuzzy inference is not precise, but is approximate. However, the inference does represent an approximate linguistic characteristic of the relation between two universes of discourse, X and Y. Other works in approximate reasoning can be found in Zadeh, (1973), Mamdani (1976), Mizumoto and Zimmerman (1982), and Yager (1983, 1985).

REFERENCES

Mamdani, E. H. (1976) "Advances in linguistic synthesis of fuzzy controllers," *Int. J. of Man-Machine Studies,* **8**, 669-678.

Mizumoto, M. and Zimmerman, H.-J. (1982) "Comparison of fuzzy reasoning methods," *Fuzzy Sets and Systems,* **8**, 253-283.

Yager, R. R. (1983) "On the implication operator in fuzzy logic," *Information Sciences,* **31**, 141-164.

Yager, R. R. (1985) "Strong truth and rules of inference in fuzzy logic and approximate reasoning," *Cybernetics and Systems,* **16**, 23-63.

Zadeh, L. A. (1973) "Outline of a new approach to the analysis of complex systems and decision processes," *IEEE on Trans. Systems, Man and Cybernetics,* **SMC-1**, 28-44.

4

FUZZY RULE-BASED EXPERT SYSTEMS I

Nader Vadiee
University of New Mexico

This chapter originally appeared in the journal *Intelligent and Fuzzy Systems: Applications in Engineering and Technology*, March 1993. Used here by permission.

The general problems of system identification and control system design are formulated and the concept of control or decision surface is described. It is shown that a collection of fuzzy conditional restrictive rules could be used to model complex or ill-defined processes (fuzzy systems), and some simple canonical formats for the production rule-sets are given. It is shown that a system of fuzzy relational equations could be obtained from a set of IF-THEN production rules. A fuzzy system transfer relation based on a system of fuzzy relational equations is defined. It is also described how a variety of fuzzy implication relations or Zadeh's extension principle might be used to derive the fuzzy relational equations. Various methods for obtaining the solution of these equations based on a number of composition of relations techniques are studied. Finally, two most commonly used solution techniques in fuzzy control applications, i.e., Max.-Min. and Max.-Product methods, are described.

4.1   INTRODUCTION

A mathematical model that describes a wide variety of physical systems is an nth-order ordinary differential equation of the type:

$$\frac{d^n y(t)}{dt^n} = w\left[t, y(t), \dot{y}(t), \cdots, \frac{d^{n-1} y(t)}{dt^{n-1}}, u(t)\right] \tag{1}$$

where t is the time parameter, u(.) is the input function, and y(.) is the output or response function. If we define the auxiliary functions:

$$x_1(t) = y(t)$$

$$x_2(t) = \dot{y}(t)$$

$$\vdots$$

$$x_n(t) = \frac{d^{n-1} y(t)}{dt^{n-1}} \tag{2}$$

then the single nth-order equation (1) can be equivalently expressed as a system of n first-order equations:

$$\dot{x}_1(t) = x_2(t)$$

$$\dot{x}_2(t) = x_3(t)$$

$$\vdots$$

$$\dot{x}_{n-1}(t) = x_n(t) \tag{3}$$

Finally, if we define n-vector-valued functions x(.) and f(.) by

$$\mathbf{x}(t) = [x_1(t), x_2(t), \cdots, x_n(t)]' \tag{4}$$

$$\mathbf{f}(t, \mathbf{x}, u) = [x_2, x_3, \cdots, x_n, w(t, x_1, \cdots, x_n, u)]' \tag{5}$$

where x(t) is the system state vector at time t, then the n first-order equations (3) can be combined into a first-order vector differential equation, i.e.,

$$\dot{\mathbf{x}}(t) = \mathbf{f}[t, \mathbf{x}(t), u(t)] \tag{6}$$

and the output y(t) is given from (2) as:

$$y(t) = [1, 0, \cdots, 0]\, \mathbf{x}(t) \tag{7}$$

Chap. 4 Fuzzy Rule-Based Expert Systems-I

Similarly, a system with p inputs, m outputs, and n states, will be described, in general, as:

$$\dot{\mathbf{x}}(t) = \mathbf{f}\ [t, \mathbf{x}(t), \mathbf{u}(t)] \tag{8}$$

$$\mathbf{y}(t) = \mathbf{g}\ [t, \mathbf{x}(t), \mathbf{u}(t)] \tag{9}$$

where $\mathbf{u}(t)$ and $\mathbf{y}(t)$ vectors defined as:

$$\mathbf{u}(t) = [u_1(t), u_2(t), \cdots, u_p(t)]' \tag{10}$$

$$\mathbf{y}(t) = [y_1(t), y_2(t), \cdots, y_m(t)]' \tag{11}$$

are input vector and output vector, respectively. Physical systems descriptions based on equations (8) and (9) are known as state-space representations. In the case of time-invariant systems, equations (8) and (9) become:

$$\dot{\mathbf{x}}(t) = \mathbf{f}\ [\mathbf{x}(t), \mathbf{u}(t)] \tag{12}$$

$$\mathbf{y}(t) = \mathbf{g}\ [\mathbf{x}(t), \mathbf{u}(t)] \tag{13}$$

and for a linear time-invariant system

$$\dot{\mathbf{x}}(t) = A \cdot \mathbf{x}(t) + B \cdot \mathbf{u}(t) \tag{14}$$

$$\mathbf{y}(t) = C \cdot \mathbf{x}(t) + D \cdot \mathbf{u}(t) \tag{15}$$

where constants A, B, C, and D are known as system matrices.

A first-order single-input and single-output system is described using a discrete-time equation as:

$$x_{k+1} = f(x_k, u_k) \tag{16}$$

where x_{k+1}, x_k are the values of state at time moments kth and $(k+1)$th and u_k is the input at the kth moment. An nth order single-input and single-output system can be put in the form of:

$$y_{k+n} = f(y_k, y_{k+1}, \cdots, y_{k+n-1}, u_k) \tag{17}$$

and for an nth-order, multiple-input and single-output discrete system:

$$y(k + n) = f[(y(k), y(k + 1), \cdots, y(k + n - 1), u_1(k), u_2(k), \cdots, u_p(k)] \tag{18}$$

4.1.1 System Identification Problem

The general problem of identifying a physical system based on the measurements of the input, output, and state variables is defined as obtaining functions **f** and **g**, in the case of a nonlinear system, and system matrices A, B, C, and D, in the case of a linear system. There exist algorithms that adaptively converge to these system parameters based on numerical data taken from input and output variables. Fuzzy systems and artificial neural network paradigms are two evolving disciplines for nonlinear system identification problems.

4.1.2 Control System Design Problem

The general problem of feedback control system design is defined as obtaining a generally nonlinear vector-valued function **h**(.) defined as:

$$\mathbf{u}(t) = \mathbf{h}[t, \mathbf{x}(t), \mathbf{r}(t)] \tag{19}$$

where **u**(t) is the input to the plant or process, **r**(t) is the reference input, and **x**(t) is the state vector. The feedback control law **h** is supposed to stabilize the feedback control system and result in a satisfactory performance.

In the case of a time-invariant system with a regulatory type of controller, the control command could be stated as either a state-feedback or an output feedback as shown in the following:

$$\mathbf{u}(t) = \mathbf{h}[\mathbf{x}(t)] \tag{20}$$

$$\mathbf{u}(t) = \mathbf{h}[y(t), \dot{y}, \int y dt] \tag{21}$$

In the case of a simple single-input and single-output system the function **h** takes one of the following forms:

$$\mathbf{u}(t) = K_p \cdot e(t)$$

for a proportional or **P** controller;

$$\mathbf{u}(t) = K_p \cdot e(t) + K_I \cdot \int e(t) dt$$

for a proportional plus integral or PI controller;

$$\mathbf{u}(t) = K_p \cdot e(t) + K_D \cdot \dot{e}(t)$$

Chap. 4 Fuzzy Rule-Based Expert Systems-I

for a proportional plus derivative or PD controller;

$$u(t) = K_p \cdot e(t) + K_I \cdot \int e(t)dt + K_D \cdot \dot{e}(t)$$

for a proportional plus derivative plus integral or PID controller, where $e(t)$, $\dot{e}(t)$, and $\int e(t)dt$ are the output error, error derivative, and error integral, respectively; and

$$u(t) = -[k_1 \cdot x_1(t) + k_2 \cdot x_2(t) + \cdots + k_n \cdot x_n(t)]$$

for a full state-feedback controller.

The problem of control system design is defined as obtaining the generally nonlinear function $h(.)$ in the case of nonlinear systems, coefficients K_P, K_I, and K_D in the case of an output-feedback, and coefficients $k_1, k_2, \ldots,$ and k_n, in the case of state-feedback policies for linear system models.

4.1.3 Control Surface

The function h as defined in equations (19), (20), and (21) is, in general, defining a nonlinear hypersurface in an n-dimensional space. For the case of linear systems with output feedback or state feedback it is a hyperplane in an n dimensional space. This surface is known as the control or decision surface.

The control surface describes the dynamics of the controller and is generally a time-varying non-linear surface. Due to unmodeled dynamics present in the design of any controller, techniques should exist for adaptively tuning and modifying the control surface shape. Fuzzy logic rule-based expert systems use a collection of fuzzy conditional statements derived from an expert knowledge-base to approximate and construct the control surface. This paradigm of control system design is based on interpolative and approximate reasoning. Fuzzy rule-based controllers or system identifiers are generally model-free paradigms. Fuzzy logic rule-based expert systems are universal nonlinear function approximators and any nonlinear function of n independent variables and one dependent variable can be approximated to any desired precision.

On the other hand, artificial neural networks are based on analogical learning and try to learn the nonlinear function through adaptive and converging techniques and based on numerical data available from input-output measurements on the system variables and some performance criteria.

4.1.4 Control System Design Stages

The seven basic steps in designing a complex or ill-defined physical system are:

i) Large scale systems are decentralized and decomposed into a collection of decoupled sub-systems.

ii) The temporal variations of plant dynamics are assumed to be "slowly varying."

iii) The nonlinear plant dynamics are locally linearized about a set of operating points.

iv) A set of state variables, control variables, or output features are made available.

v) A simple P, PD, PID (output feedback), or state-feedback controller is designed for each de coupled system. The controllers are of regulatory type and are fast enough to perform satisfactorily under tracking control situations. Optimal controllers can also be tried using LQR or LQG techniques.

vi) The first five steps mentioned above introduce uncertainties. There are also uncertainties due to external environment. The controller design should be made as close as possible to the optimal one based on the control engineer's all best available knowledge, in the form of I/O numerical observations data, analytic, linguistic, intuitive, and etc., information regarding the plant dynamics and external world.

vii) A supervisory control system, either automatic or a human expert operator, forms an additional feedback control loop to tune and adjust the controller's parameters in order to compensate the effects of uncertainties and variations due to unmodeled dynamics.

4.1.5 Assumptions in a Fuzzy Control System Design

Six basic assumptions are commonly made whenever a fuzzy logic-based control policy is selected. These assumptions are outlined below:

i) The plant is observable and controllable: State, input, and output variables are available for observation and measurement or computation.

ii) There exists a body of knowledge in the form of expert production linguistic rules, engineering common sense, intuition, or an analytic model that can be fuzzified and the rules be extracted.

iii) A solution exists.

vi) The control engineer is looking for a good enough solution and not necessarily the optimum one.

v) We desire to design a controller to the best of our available knowledge and within an acceptable precision range.

vi) The problems of stability and optimality are open problems.

Chap. 4 Fuzzy Rule-Based Expert Systems-I

The following sections discuss the problem of obtaining the control surface $h(.)$, i.e., approximations based on a collection of fuzzy IF-THEN rules which describe the dynamics of the controller. Fuzzy rule-based expert models can also be used to obtain acceptable approximations for the functions $f(.)$ and $g(.)$ in the case of a system identification problem.

A fuzzy production rule system consists of four structures:

i) A set of rules which represents the policies and heuristic strategies of the expert decision-maker.

ii) A set of input data assessed immediately prior to the actual decision.

iii) A method for evaluating any proposed action in terms of its conformity to the expressed rules, given the available data.

iv) A method for generating promising actions and for determining when to stop searching for better ones.

The input data, rules, and output action or consequence are generally fuzzy sets expressed by means of appropriate membership functions defined on a proper space. The method of evaluation of rules is known as *approximate reasoning* or *interpolative reasoning*, and is commonly represented by composition of fuzzy relations applied to a fuzzy relational equation.

4.2 FUZZY RULE-BASED EXPERT SYSTEMS (CANONICAL FORMS)

Consider an n-input and m-output system shown in Figure 4-1. Let X be a Cartesian product of n universes x_i, for $i = 1, 2, \ldots, n$, i.e., $X = X_1 \times X_2 \times \cdots \times X_n$; and Y be a Cartesian product of m universes y_j for $j = 1, 2, \ldots, m$, i.e., $Y = Y_1 \times Y_2 \times \cdots \times Y_m$. $x = (x_1, x_2, \ldots, x_n)'$ is the input vector to the system defined on real space R^n and $y = (y_1, y_2, \ldots, y_m)'$ is the output vector of the system defined on real space R^m. The system S could represent any general static nonlinear ng from X to Y, an industrial control system, a dynamic system identification mapping model, a pattern recognition system, or a decision-making process.

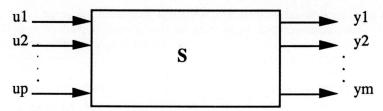

Figure 4-1. A block diagram for a p-input and m-output system.

There are three general spaces present in all these systems and processes, i.e.,

i) the space of possible conditions at the inputs to the system and, in general, could be represented by a collection of fuzzy subsets A^K, for k= 1, 2, ..., which are fuzzy partitions of space **X**, expressed by means of membership functions

$$A^k \varepsilon\ F(X);\ \mu_A k(x): X \to [0, 1], \text{ for } k = 1, 2, \cdots \qquad (22)$$

ii) the space of possible output consequences, control commands from a controller, or the decisions recommended to or made by a decision-maker, based on some specific conditions at the inputs, and, in general, could be represented by a collection of fuzzy subsets B^K, for k = 1, 2, ..., which are fuzzy partitions of space **Y**, expressed by means of membership functions

$$B^k \varepsilon\ F(Y);\ \mu_B k(y): Y \to [0, 1], \text{ for } k = 1, 2, \cdots \qquad (23)$$

and,

iii) the space of possible mapping relations from the input space **X** onto the output space **Y**. The mapping relations are, in general, represented by fuzzy relations R^K, for k = 1, 2, ..., and expressed by means of membership functions

$$R^k \varepsilon\ F(X \times Y);\ \mu_R k(x, y): X \times Y \to [0, 1], \text{ for } k = 1, 2, \cdots \qquad (24)$$

where F(.), in equations (22), (23), and (24), denotes a family of fuzzy sets on a proper space. A human perception of the system S is based on experience and expertise, intuition, a knowledge of the physics of the system, or a set of subjective preferences and goals. This type of knowledge is usually put in the form of a set of unconditional as well as conditional propositions in natural language, by the human expert. Our understanding of complex humanistic and non-humanistic systems is at a qualitative and declarative level, based on vague linguistic terms. This is called a fuzzy level of understanding of a physical system.

The unconditional as well as conditional statements, in general, place some restrictions on the consequent of the process based on certain immediate as well as past conditions. These restrictions are usually vague natural language terms and words that could be modeled using fuzzy mathematics.

Consider the problem of the coloring of a landscape and some expert restrictive statements:

Chap. 4   Fuzzy Rule-Based Expert Systems-I

If the season is spring, then the color is rather light green.
If the season is summer, then the color is deep green.
If the season is fall, then the color is bright and deep yellow.
.
.
.
, etc.

The vague term "rather light green" in the first statement places a fuzzy restriction on the color, based on a fuzzy "spring" condition in the antecedent. A similar case is a pattern recognition problem such as "If the color is red, the tomato is ripe." A room temperature control problem might contain the following expert rules:

If the room temperature is very hot,
   then
      If the heat is on
         then turn the heat lower.

In this case restrictions are placed on the actions taken by a controller, i.e., "turn the heat lower."

In summary, the fuzzy level of understanding and describing a complex system basically is put in the form of a set of restrictions on the output based on certain conditions at the input. Restrictions are generally modeled by fuzzy sets and relations.

4.2.1   Canonical Fuzzy Rule-Based Expert System

The canonical form of a set of expert production rules is defined as a set of unconditional restrictions followed by a set of conditional restrictions. These restriction statements, unconditional as well as conditional, are usually connected by linguistic connectives such as "and," "or," or "else." The canonical form of an expert production rule-set is given in Figure 4-2.

R^1	Restriction R^1
R^2	Restriction R^2
⋮	
R^k	Restriction R^k
R^{k+1}	IF condition C^1, THEN restriction R^{k+1}
R^{k+2}	IF condition C^2, THEN restriction R^{k+2}
⋮	
R^r	IF condition C^{r-k}, THEN restriction R^r

Figure 4-2. The canonical form of a set of expert rules.

The restrictions $R^1, R^2, \ldots, R^r$ apply to the output actions taken on the output, or the decisions made for a desired performance.

In general, there exist three general statement forms in any linguistic algorithm or expert production rule-set.

    i)   Assignment statements, e.g.:

        $x \cong s$
        x = small
        season = spring
        room temperature = hot
        tomato's color = red
        x is large
        x is not large and not very small

    ii)   Conditional statements, e.g.:
        IF x is small THEN y is large ELSE y is not large
        IF x is positive THEN decrease y slightly
        IF the tomato is red THEN the tomato is ripe
        IF x is very small THEN stop

    iii)   Unconditional action statements, e.g.:
        multiply by x
        turn the heat lower
        delete first few terms
        go to 7
        stop

Chap. 4   Fuzzy Rule-Based Expert Systems-I   61

The unconditional propositions, equivalently, may be thought of as conditional restrictions with their IF clause condition being the universe of discourse of the input conditions, which is always true and satisfied. An unconditional restriction such as "output is low" could, equivalently, be written as:

IF any conditions THEN output is low

or

IF anything THEN low.

Hence, the system under consideration could be described using a collection of conditional restrictive statements. These statements may also be modeled as fuzzy conditional statements, such as:

IF condition C^1 THEN restriction R^1.

The unconditional restrictions might be in the form:

R^1:   The output is B^1
         AND
R^2:   The output is B^2
         AND
.
.
.

where $B^1, B^2, \ldots$ are fuzzy subsets defined in equation (23).

Figure 4-3 is the expert system comprised of a set of conditional rules. Hence, the canonical rule set may be put in the following form:

Rule 1:   IF condition C^1, THEN restriction R^1

Rule 2:   IF condition C^2, THEN restriction R^2
.
.
.
Rule R^r IF condition C^r, THEN restriction R^r

Figure 4-3. The canonical fuzzy rule-based expert system.

For the case of n-input and m-output system S, described earlier, the canonical fuzzy rule-based expert system (FRBES) could be put in the form shown in Figure 4-4.

> Rule 1: IF x is A^1, THEN y is B^1
>
> Rule 2: IF x is A^2, THEN y is B^2
>
> .
> .
> .
>
> Rule r: IF x is A^r, THEN y is B^r

Figure 4-4. FRBES describing the system S.

and in a more compact form:

(IF A^1 THEN B^1 "α" IF A^2 THEN B^2 "α" ... "α" IF A^r THEN B^r)

where "α" could be any of "and," "or," or "else" linguistic connectives. In Figure 4-5, **x** and **y** are the input and output vectors, respectively. The input to the system could have crisp and sharp as well as fuzzy values. It is noted that crisp values $x = \overline{x}$, or $y = \overline{y}$, known as fuzzy singletons, are expressed by a membership function as follows:

$$\mu_A(x) = \delta_{x,\overline{x}} = \delta(x - \overline{x}) = \begin{cases} 1, & \text{for } x = \overline{x} \\ 0, & \text{otherwise} \end{cases} \qquad (25)$$

and, similarly,

$$\mu_B(y) = \delta_{y,\overline{y}} = \delta(y - \overline{y}) = \begin{cases} 1, & \text{for } y = \overline{y} \\ 0, & \text{otherwise} \end{cases} \qquad (26)$$

or a crisp interval defined as $a \leq x \leq b$, or $c \leq y \leq d$ may be defined as fuzzy sets expressed by the membership functions

$$\mu_A(x) = \begin{cases} 1, & \text{for } a \leq x \leq b \\ 0, & \text{otherwise} \end{cases} \qquad (27)$$

and

$$\mu_B(y) = \begin{cases} 1, & \text{for } c \leq y \leq d \\ 0, & \text{otherwise} \end{cases} \qquad (28)$$

Therefore, crisp and fuzzy conditions and consequences could be treated in a common setting.

Chap. 4   Fuzzy Rule-Based Expert Systems-I   63

4.2.2 Decomposition of Compound Rules into Simple Canonical Forms

In a linguistically-expressed algorithmic or descriptive form given by a human expert, more compound rule structures could be present. As an example, consider an algorithm for the control of temperature in a room:

    IF it is raining <u>hard</u>
        THEN close the window.
    IF the room temperature is very hot,
        THEN
            IF the heat is on
                THEN turn the heat lower
            ELSE
                IF (the window is closed) AND (the air
                      conditioner is off)

        AND (it is not raining hard)
            THEN open the window
        ELSE
            IF (the window is closed)
            AND (the air conditioner is on)
                THEN open the window.

    .
    .
    .

    IF "the temperature is cold" is fairly true
        THEN
            IF the air conditioner is on
                THEN turn the air conditioner off.
    etc.

4.2.3 Basic Linguistic Terms

In general, a value of a linguistic variable is a composite term which is a concatenation of atomic terms. These atomic terms may be divided into four categories:

    i)   primary terms which are labels of specified fuzzy subsets of the universe of discourse (e.g., hot, cold, hard, lower, etc., in the preceding example).

    ii)   The negation NOT and connectives "AND" and "OR."

    iii) Hedges, such as "very," "much," "slightly," "more or less," etc.

    iv) Markers, such as parentheses.

The primary as well as composite terms may also be followed by linguistic variable labeled likelihood, such as "likely," "very likely," "highly likely," "unlikely," etc. In addition, primary and composite terms might also be modified semantically by truth qualification statements such as "true," "fairly true," "very true," "false," fairly false," and "very false." Likelihood labels are based on probability. The primary terms, as well as the rules, may also be restricted by the linguistic variable labeled certainty, such as "indefinite," "unknown," and "definite." The conditional rules might as well be simple or compound. Compound conditional rules may be in the form of nested IF-THEN rules or rules with linguistic terms such as "unless" or "else."

Figure 4-4 depicts the basic canonical form of FRBES that we deal with. Based on basic properties and operations defined for fuzzy sets, it can be shown that any compound rule structure may be decomposed and reduced to a number of simple canonical rules as given in Figure 4-4. These rules are based on natural language representations and models which are themselves based on fuzzy sets and fuzzy logic. The following illustrates a number of the most common techniques for decomposition of compound linguistic rules.

4.2.4 Primary Terms Preceded by Linguistic Hedges

In general, the linguistic hedges such as "very," "more or less," "slightly," "sort of," "more than," "less than," "essentially," etc., whenever operating on a primary fuzzy term, are equivalent to a specified nonlinear transformation of the membership function of the primary term into a new membership function of the composite term.

Consider a universe of discourse on the set of integer numbers in the interval [1, 5]. The fuzzy subset "small" may be expressed by means of membership function:

$$\text{"small"} = \frac{1}{1} + \frac{0.8}{2} + \frac{0.6}{3} + \frac{0.4}{4} + \frac{0.2}{5}$$

and the fuzzy subset "large" may be defined as

$$\text{"large"} = \frac{0.2}{1} + \frac{0.4}{2} + \frac{0.6}{3} + \frac{0.8}{4} + \frac{1}{5}$$

"Very small" could then be defined as a concentration of "small," or

$$\text{"very small"} = (\text{small})^2$$

$$= \int \frac{[m_{small}(x)]^2}{x}$$

$$= \frac{1}{1} + \frac{0.64}{2} + \frac{0.36}{3} + \frac{0.16}{4} + \frac{0.04}{5}$$

and in the same manner:

Chap. 4 Fuzzy Rule-Based Expert Systems-I

"very, very small" $= (\text{small})^4 = (\text{very small})^2$

$$= \int \frac{[\mu_{\text{very small}}(x)]^2}{x}$$

$$= \frac{1}{1} + \frac{0.4}{2} + \frac{0.1}{3}$$

where small terms have been neglected.

Based on the definition of fuzzy complement operation, "not very small" may be defined as:

$$= \int \frac{1 - \mu_{\text{very small}}(x)}{x}$$

$$\cong \frac{0.4}{2} + \frac{0.6}{3} + \frac{0.8}{4} + \frac{1}{5}$$

As a more complicated example, consider the composite term

x = not very small and not very large, where "very large" is defined as a concatenation of "very" and "large," as:

very large $= (\text{large})^2$

$$= \frac{0.04}{1} + \frac{0.16}{2} + \frac{0.36}{3} + \frac{0.64}{4} + \frac{1}{5}$$

and "not very large" as:

not very large $= \text{NOT (very large)}$

$$\cong \frac{1}{1} + \frac{0.8}{2} + \frac{0.6}{3} + \frac{0.4}{4}$$

Based on fuzzy intersection operation for "AND," we define the fuzzy subset "not very small and not very large" as:

not very small and not very large
$= (\text{not very small}) \cap (\text{not very large})$

$$= \left(\frac{0.4}{2} + \frac{0.6}{3} + \frac{0.8}{4} + \frac{0.96}{5}\right) \cap \left(\frac{1}{1} + \frac{0.9}{2} + \frac{0.6}{3} + \frac{0.4}{4}\right)$$

$$= \frac{0.4}{2} + \frac{0.6}{3} + \frac{0.4}{4}$$

Here are some other linguistic hedge operations:

"plus A" $\triangleq A^{1.25}$

"minus A" $\triangleq A^{0.75}$

"more than A" $\triangleq \begin{cases} 0, & \text{for } x \leq x_0 \\ 1 - \mu_A(x), & \text{for } x > x_0 \end{cases}$

where x_0 is the element attaining the maximum grade of membership in A, and likewise:

"less than A" $\triangleq \begin{cases} 0, & \text{for } x \geq x_0 \\ 1 - \mu_A(x), & \text{for } x < x_0 \end{cases}$

and,

"plus plus A" = minus very A

"highly A" = minus very very A

and equivalently:

"highly A" = plus plus very A

4.2.5 Likelihood Linguistic Labels

An example of a different nature is provided by the values of a linguistic variable labeled likelihood. In this case, we assume that the universe of discourse is given by

$$U = \{0, 0.1, 0.2, 0.3, 0.4, 0.5, 0.6, 0.7, 0.8, 0.9, 1.0\}$$

in which the elements of U represent probabilities.

Suppose that we wish to compute the meaning of the value

x = highly unlikely

in which "highly" is defined as

Chap. 4 Fuzzy Rule-Based Expert Systems-I

and
> highly = minus very very
>
> unlikely = not likely.

With the meaning of the primary term "likely" given by

$$\text{likely} = \frac{1}{1} + \frac{1}{0.9} + \frac{1}{0.8} + \frac{0.8}{0.7} + \frac{0.6}{0.6} + \frac{0.5}{0.5} + \frac{0.3}{0.4} + \frac{0.2}{0.3}$$

we obtain

$$\text{unlikely} = \frac{1}{0} + \frac{1}{0.1} + \frac{1}{0.2} + \frac{0.8}{0.3} + \frac{0.7}{0.4} + \frac{0.5}{0.5} + \frac{0.4}{0.6} + \frac{0.2}{0.7}$$

and hence

$$\text{very very unlikely} = (\text{unlikely})^4$$

$$\cong \frac{1}{0} + \frac{1}{0.1} + \frac{1}{0.2} + \frac{0.4}{0.3} + \frac{0.2}{0.4}.$$

Finally,

$$\text{highly unlikely} = \text{minus very very unlikely}$$

$$\cong \left(\frac{1}{0} + \frac{1}{0.1} + \frac{1}{0.2} + \frac{0.4}{0.3} + \frac{0.2}{0.4}\right)^{0.75}$$

$$\cong \frac{1}{0} + \frac{1}{0.1} + \frac{1}{0.2} + \frac{0.5}{0.3} + \frac{0.3}{0.4}.$$

The primary terms "yes," "maybe," and "no" may also be assigned meanings based on membership functions given by "very very likely," "likely," and "very very unlikely."

It is also noted that the atomic term "anything" is equivalent to the universe of discourse and given by

$$\mu_{\text{anything}}(x) = 1; \text{ for all } x \in X$$

It should be noted that in computing the meaning of the composite terms in the preceding examples we have made implicit use of the usual precedence rules governing the evaluation of Boolean expressions. With the addition of hedges, these precedence rules may be expressed as follows:

Precedence	Operation
First	hedges, not
Second	and
Third	or

As usual, parentheses may be used to change the precedence order and ambiguities may be resolved by the use of association to the right. Thus "plus very minus very tall" should be interpreted as

    plus (very(minus(very(tall))))

In summary, every atomic term as well as every composite term has a syntax represented by its linguistic label and a semantics or meaning which is given by a membership function. The concept of membership function gives an elastic and flexible meaning to a linguistic term. Based on this elasticity and flexibility, it is possible to incorporate subjectivity and bias into the meaning of a linguistic term. This is one of the most important benefits of fuzzy mathematics introduction into the modeling of linguistic propositions. Another important benefit is the straightforward and simple computational tools for processing and manipulating the meaning of composite linguistic terms and compound linguistic propositions. These capabilities and tools of fuzzy mathematics are used to encode and automate human expert knowledge, which is often expressed by natural language propositions, for implementation on computers.

    The subjectivity that exists in fuzzy modeling is a blessing rather than a curse. The subjectivity present in the definition of terms is balanced by the subjectivity in the conditional rules used by an expert. As far as the set of variables and their meanings are compatible and consistent with the set of conditional rules used, the overall outcome turns out to be objective and meaningful. Fuzzy mathematical tools and the calculus of fuzzy IF-THEN rules opened the way for automation and implementation of a huge body of human expert knowledge. It provided a means of sharing, communicating, or transferring of human expert subjective knowledge of systems and processes.

4.2.6   Truth Qualification Rate

Let τ be a fuzzy truth value, for example, "very true," "true," "fairly true," "fairly false," "false," etc. Such a truth value may be regarded as a fuzzy subject of the unit interval that is characterized by a membership function $x_t : [0, 1] \rightarrow [0 \times 1]$. A truth qualification proposition can be expressed as "x is A is τ." The translation rule for such propositions can be given by:

$$x \text{ is } A \text{ is } \tau \rightarrow \mu = \mu_x+ \quad (29)$$

$$\mu_x+(x) = x_\tau[\mu_x(x)] \quad (30)$$

Figure 4-5 illustrates the meaning of true (T), fairly true (FT), very true (VT), false (F), fairly false (FF), and very false (VF). The fuzzy assignment statements such as:

    x is A is very true

    y is B is very false

and fuzzy conditional statements such as

Chap. 4   Fuzzy Rule-Based Expert Systems-I

It is very true that IF A THEN B

are transformed to a new meaning using equations (29) and (30). In the case of conditional statements, the fuzzy set describing the fuzzy relation A → B will be transformed to a new fuzzy relation.

As an example: if x has a membership value equal to 0.85 in the fuzzy set A, then its membership values of:

        x is A   is true
        x is A   is false
        x is A   is fairly true
        x is A   is very false

are 0.85, 0.15, 0.92 and 0.22, respectively, which clearly is based on Figure 4-5.

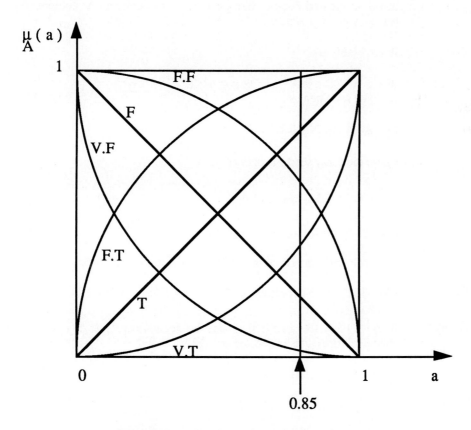

Figure 4-5. Truth value qualification graphs.

4.2.7 Multiple Antecedents or Consequents Connected by "AND" or "OR"

The following discussion covers a number of general compound rules structures with more than one antecedent or consequent and illustrates methods for transforming these compound rules into simple canonical forms.

i) IF x is A^1 and A^2 ... and A^L THEN y is B^S
Assuming a new fuzzy subset A^S as

$$A^S = A^1 \cap A^2 \cap \cdots \cap A^L$$

expressed by means of membership function

$$\mu_{A^S}(x) = \text{Min}[\mu_{A^1}(x), \mu_{A^2}(x), \cdots, \mu_{A^L}(x)]$$

based on the definition of fuzzy intersection operation, the compound rule may be rewritten as

IF A^S THEN B^S

ii) IF x is A^1 OR x is A^2 ... OR x is A^L THEN y is B^S
could be rewritten as

IF x is A^S THEN y is B^S

where the fuzzy set A^S is defined as

$$A^S = A^1 \cup A^2 \cup \cdots \cup A^L$$

$$\mu_{A^S}(x) = \text{Max}[\mu_{A^1}(x), \mu_{A^2}(x), \cdots, \mu_{A^L}(x)]$$

which clearly is based on the definition of fuzzy union operation.

4.2.8 Conditional Statements With "ELSE," "UNLESS"

i) IF A^1 THEN (B^1 ELSE B^2) may be decomposed into two simple canonical form rules connected by "OR":

OR
    IF A^1 THEN B^1
    IF NOT A^1 THEN B^2

ii) IF A^1 (THEN B^1) UNLESS A^2 could be decomposed as

Chap. 4 Fuzzy Rule-Based Expert Systems-I

OR
$$\text{IF } A^1 \text{ THEN } B^1$$
$$\text{IF } A^2 \text{ THEN NOT } B^1$$

iii) IF A^1 THEN (B^1 ELSE IF A^2 THEN (B^2)) may be put into the following form:

OR
$$\text{IF } A^1 \text{ THEN } B^1$$
$$\text{IF NOT } A^1 \text{ AND } A^2 \text{ THEN } B^2$$

4.2.9 Nested IF-THEN Rules

i) IF A^1 THEN (IF A^2 THEN (B^1)) may be put into the form:

$$\text{IF } A^1 \text{ AND } A^2 \text{ THEN } B^1$$

ii) IF A^1 THEN (IF A^2 THEN (IF ... (B^1) ...) is the general case which is rewritten as:

$$\text{IF } A^1 \text{ AND } A^2 \text{ AND ... THEN } B^1$$

iii) IF A^1
    THEN
        IF A^2
        THEN
            IF A^3
            THEN B^1 ELSE B^2
ELSE B^3
may be transformed to:

OR
$$\text{IF } A^1 \text{ AND } A^2 \text{ AND } A^3 \text{ THEN } B^1$$
OR
$$\text{IF } A^1 \text{ AND } A^2 \text{ AND NOT } A^3 \text{ THEN } B^2$$
$$\text{IF NOT } A^1 \text{ THEN } B^3$$

Using parentheses this compound rule might also be written as:

$$\text{IF } A^1 \text{ (THEN (IF } A^2 \text{ (THEN IF } A^3 \text{ THEN } (B^1 \text{ ELSE } B^2)) \text{ ELSE } B^3)$$

As another example of nested IF-THEN rule structures, consider the following compound rule:

IF A^1
    THEN B^1 AND B^2
       IF A^2
          THEN B^3
             IF A^3
                THEN B^4

which may be put into the following form:

IF A^1 THEN B^1

IF A^1 THEN B^2

IF A^1 AND A^2 THEN B^3

IF A^1 AND A^2 AND A^3 THEN B^4

4.2.10 Canonical FRBES Forms for Multiple-input Multiple-output Physical Systems

In the following we discuss two more common canonical fuzzy rule-based expert systems.

    i) For the n-output and m-output system described earlier, if it can be assumed that input fuzzy sets $A^1, A^2, \ldots$, are comprised of n non-interactive fuzzy sets defined on universes x_i for $i = 1, 2, \ldots, n$ and also that the output fuzzy sets $B^1, B^2, \ldots$, are composed of m non-interactive fuzzy sets defined as universes y_j for $j = 1, 2, \ldots, m$, then the canonical rule-based expert system given in Figure 4-4 could be put into the canonical form shown in Figure 4-4.

R^1:   IF x_1 is A_1^1 AND x_2 is A_2^1 ... AND x_n is A_n^1
           THEN y_1 is B_1^1 AND y_2 is B_2^1 ... AND y_m is B_m^1

R^2:   IF x_1 is A_1^2 AND x_2 is A_2^2 ... AND x_n is A_n^2
           THEN y_1 is B_1^2 AND y_2 is B_2^2 ... AND y_m is B_m^2.

       .
       .
       .

R^r: IF x_1 is A_1^r AND x_2 is A_2^r ... AND x_n is A_n^r
           THEN y_1 is B_1^r AND y_2 is B_2^r ... AND y_m is B_m^r.

Figure 4-6. Canonical FRBES for multi-input multi-output system.

Chap. 4   Fuzzy Rule-Based Expert Systems-I

In Figure 4-6, the fuzzy sets A_i^k ($i = 1, 2, \ldots n$ and $k = 1, 2, \ldots, r$) and fuzzy sets B_j^k ($j = 1, 2, \ldots, m$ and $k = 1, 2, \ldots, r$) are expressed as

$$A_i^K \in F(x_i); \; \mu_{A_i^K}(x_i) : x_i \to [0, 1] \tag{31}$$

and

$$B_j^k \in F(y_j); \; \mu_{B_j^k}(y_j) : y_j \to [0, 1] \tag{32}$$

   ii)   Systems with more than one output, e.g., *m* outputs, may be described by a collection of FRBES, each rule set dealing with only one output. Hence, the canonical FRBES describing an *n*-input and single-output system, with non-interactive input fuzzy sets, could be put in the form given in Figure 4-7.

R^1:   IF x_1 is A_1^1 and x_2 is A_2^1 ... AND x_n is A_n^1
         THEN y is B^1

R^2:   IF x_1 is A_1^2 and x_2 is A_2^2 ... AND x_n is A_n^2
         THEN y is B^2.

.
.
.

R^r: IF x_1 is A_1^r and x_2 is A_2^r ... AND x_n is A_n^r
         THEN y is B^r.

Figure 4-7. FRBES for an *n*-input and single-output system.

The fuzzy rule-based expert system given in Figure 4-7 is the most common canonical form in system identification and control problems.

4.3   SYSTEMS OF FUZZY RELATIONAL EQUATIONS

Fuzzy conditional proposition IF **A** THEN **B** is known as the *generalized modus ponen*. There are numerous techniques for obtaining a fuzzy relation R which will represent the generalized modus ponen in the form of a fuzzy relational equation given by

$$B = A \cdot R$$

where "•" represents a general method for composition of fuzzy relations. Some common techniques for obtaining the fuzzy relation R from the expert rules as well as other forms of knowledge regarding the system will be discussed in the next section.

Using fuzzy relational equations corresponding to each single rule, the FRBES given in Figure 4-5 may be described in the following form:

$$R^1: \quad y^1 = x \bullet R^1$$

$$R^2: \quad y^2 = x \bullet R^2$$

$$\vdots$$

$$R^r: \quad y^r = x \bullet R^r$$

Figure 4-8. System of fuzzy relational equations.

where y^k, for $k = 1, 2, \ldots, r$, is the output of the system contributed by the kth rule, and defined as:

$$y^k \in F(Y); \mu_{y^k}(y) : Y \rightarrow [0, 1], k = 1, 2, \cdots, r \qquad (33)$$

and x is the input fuzzy set to the system. Both x and y^k ($k = 1, 2, \ldots, r$) are written as unary fuzzy relations, of dimensions 1 x n and 1 x m, respectively. The unary relations, in this case, are actually similarity relations between the elements of the fuzzy set and a most typical or prototype element with membership value equal to unity. As an example, for the case where x is defined as:

$$x = \frac{0}{-3} + \frac{0.5}{-2} + \frac{0.8}{-1} + \frac{0.1}{0} + \frac{0.8}{1} + \frac{0.5}{2} + \frac{0}{3}$$

on a universe of discourse

$$U = \{-3, -2, -1, 0, +1, +2, +3 \}$$

it may be put in the form of a unary fuzzy relation as:

$$x = [0 \quad 0.5 \quad 0.8 \quad 0.1 \quad 0.8 \quad 0.5 \quad 0].$$

Similarly, the case of a crisp input $X = \overline{X} = -1$, or a fuzzy singleton, will be

$$x = [0 \quad 0 \quad 1 \quad 0 \quad 0 \quad 0 \quad 0].$$

Chap. 4   Fuzzy Rule-Based Expert Systems-I   75

The system of fuzzy relational equations given in Figure 4-8 describes a general fuzzy system. The system could be, equivalently, described by a crisp set of r fuzzy relations as:

$$R = \{R^1, R^2, \cdots, R^r\}$$

4.3.1   Aggregation of Rules

The problem of obtaining the overall output fuzzy set **y** from individual outputs contributed by individual rules or fuzzy relational equation is known as aggregation of rules problem.

Two simple extreme cases exist:

i)   *Conjunctive system of rules*: In the case of a system of rules which have to be jointly satisfied, the rules are connected by "and" connectives. In this case the aggregated output is found by the fuzzy intersection of all individual rule outputs as:

$$\mathbf{y} = (\mathbf{y}^1) \text{ AND } (\mathbf{y}^2) \text{ AND } \cdots \text{ AND } (\mathbf{y}^r)$$

or,

$$\mathbf{y} = (\mathbf{y}^1) \cap (\mathbf{y}^2) \cap \cdots \cap (\mathbf{y}^r) \tag{34}$$

which is defined by the membership function

$$m_y(y) = \text{Min } [\mu_{y^1}(y), \mu_{y^2}(y), \cdots, \mu_{y^r}(y)], \text{ for } y \in Y \tag{35}$$

The fuzzy system, in the case of system of conjunctive rules, could be described by a single aggregated fuzzy relational equation:

$$y = (x \bullet R^1) \text{ AND } (x \bullet R^2) \text{ AND } \ldots \text{ AND } (x \bullet R^r),$$

and equivalently:

$$y = x \bullet (R^1 \text{ AND } R^2 \ldots \text{ AND } R^r)$$

and finally

$$y = x \bullet R \tag{36}$$

where R is defined as

$$R = R^1 \cap R^2 \cap \cdots \cap R^r \tag{37}$$

The aggregated fuzzy relation R is called the *fuzzy system transfer relation*. For the case of a system with *n* non-interactive fuzzy inputs and single output described in Figure 4-7, the fuzzy relational equation (36) will be written in the form of

$$y = x_1 \bullet x_2 \bullet \ldots \bullet x_n \bullet R \tag{38}$$

ii) *Disjunctive system of rules*: For the case of disjunctive system of rules where the satisfaction of at least one rule is required, the rules are connected by the "OR" connectives. In this case the aggregated output is found by the fuzzy union of all individual rule contributions, as

$$\mathbf{y} = (\mathbf{y}^1) \text{ OR } (\mathbf{y}^2) \text{ OR} \cdots \text{ OR } (\mathbf{y}^r)$$

or:

$$\mathbf{y} = (\mathbf{y}^1) \cup (\mathbf{y}^2) \cup \cdots \cup (\mathbf{y}^r) \tag{39}$$

which is defined by the membership function

$$\mu_y(y) = \text{Max } [\mu_{y^1}(y), \mu_{y^2}(y), \cdots, \mu_{y^r}(y)], \text{ for } y \in Y \tag{40}$$

The fuzzy system, in the case of system of disjunctive rules, could be described by a single aggregated fuzzy relational equation as:

$$y = (x \bullet R^1) \text{ OR } (x \bullet R^2) \text{ OR } \ldots \text{ OR } (x \bullet R^r)$$

and equivalently

$$y = x \bullet (R^1 \text{ OR } R^2 \text{ OR } \ldots \text{ OR } R^r)$$

and finally

$$y = x \bullet R \tag{41}$$

where R is defined as

$$R = R^1 \cup R^2 \cup \cdots \cup R^r \tag{42}$$

The aggregated fuzzy relation, i.e., R, is called the fuzzy system transfer relation.

For the case of a system with n non-interactive fuzzy inputs and single output, described as in Figure 4-7, the fuzzy relational equation (41) will be written in the following form:

$$y = x_1 \bullet x_2 \bullet \ldots \bullet x_n \bullet R \tag{43}$$

the same as equation (37).

There is an interesting interpretation for the way that the fuzzy system transfer relation R is described. The fuzzy system transfer relation R is given by a crisp set of fuzzy relations representing the rules:

$$R = \left\{R^1, R^2, \cdots, R^r\right\}$$

and in its aggregated form is given by equations (37) and (42). Each individual relation R^k, for $k = 1, 2, \ldots, r$, represents a fuzzy data point in the Cartesian product space $X \bullet Y$. The fuzzy system transfer relation, i.e., R, is being approximated by r fuzzy input-output data point R^k, for $k = 1, 2, \ldots,$ and r. It is analogous to the case where a crisp fraction y = f(x) is approximately described by r numerical input-output values. Each $A^k \to B^k$ implication gives a fuzzy data point for approximating the overall fuzzy system transfer relation R. The more the number of these fuzzy data points with overlapping supports, a better approximation of the system input-output mapping is obtained.

The fuzzy system transfer relation R, described in this section, is a parallel computational process. The output of the fuzzy system, described by $y = x \cdot R$, is the aggregated outcome of r parallel fuzzy relations R1, R2, ..., Rr. Figure 4-9 illustrates this parallel operation where " a" represents a general aggregation method such as union (Max.) or intersection (Min.).

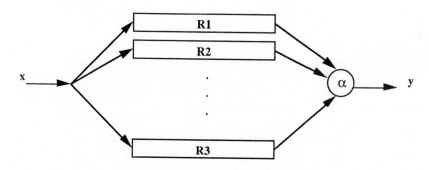

Figure 4-9. Parallel computation in FRBES models.

Fuzzy systems described by fuzzy transfer relations could be combined in series or in parallel. The corresponding equivalent fuzzy system, in the case of series connection of the fuzzy systems, could be found using composition of fuzzy relations. Figure 4-10 shows the series connection of two fuzzy systems and their corresponding equivalent system.

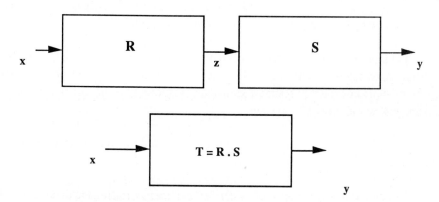

Figure 4-10. Series connection of two fuzzy systems.

Fuzzy systems in parallel connection could also be aggregated into a single equivalent fuzzy system by using an appropriate aggregation method. Figure 4-11 shows two general cases,

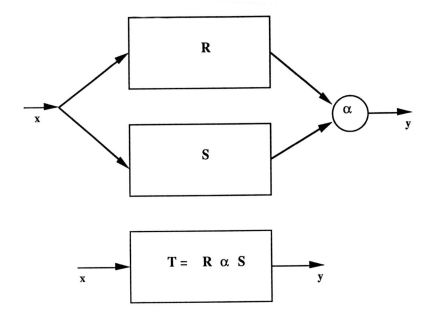

Figure 4-11. Parallel connection of fuzzy systems.

4.3.2   Obtaining the Fuzzy Relation R^k

The set of fuzzy restrictions, describing a real world dynamic system or process, is, actually, the perceptual or cognitive level of understanding and describing it. These fuzzy restrictions could be found based on one of the following methods:

i) Human expert knowledge expressed in the form of linguistic IF . . . THEN. . . rules. These rules can be extracted from interviews conducted or protocols obtained when gathering the human expert or operator knowledge regarding identification or control of an existing real world process or system.

ii) Common sense and intuitive knowledge of the design engineer about the real world process or system under investigation.

iii) Using the general physical principles and laws governing the dynamics of the process or the system under study.

Chap. 4 Fuzzy Rule-Based Expert Systems-I

iv) Using pattern classification, clustering, and statistical analysis of some available set of input-output numerical results obtained from measurements carried on an existing system or process.

v) Starting with some available closed form analytical equations which describe the process or the system and using the Zadeh's extension principles to come up with a set of fuzzy restrictions on the input-output mapping of the process or the system.

The following describes two of the most common techniques for obtaining the fuzzy relation R representing the generalized modus ponen IF **A** THEN **B**. Based on either one of these techniques a fuzzy relational equation may be found for each canonical rule of the fuzzy rule-based expert system given in Figures 4-4, 4-6, and 4-7.

i) *Fuzzy implications*: There are many different techniques for obtaining the fuzzy relation **R** based on the fuzzy sets of the IF-part and THEN-part of the fuzzy conditional proposition IF **A** THEN **B**. These are known as fuzzy implication relations. In the following we mention nine different techniques for obtaining the membership function values of fuzzy relation R defined on the Cartesian product space **X** x **Y**.

$$\mu_R(x, y) = \text{Max} \{\text{Min} [\mu_A(x), \mu_B(y)], 1 - \mu_A(x)\} \quad (44)$$

$$\mu_R(x, y) = \text{Max} \{[\mu_B(y), [1 - \mu_A(x)]\} \quad (45)$$

$$\mu_R(x, y) = \text{Min} [\mu_A(x), \mu_B(y)] \quad (46)$$

$$\mu_R(x, y) = \text{Min} \{1, [1 - \mu_A(x) + \mu_B(y)]\} \quad (47)$$

$$\mu_R(x, y) = \text{Min} \{1, [\mu_A(x) + \mu_B(y)]\} \quad (48)$$

$$\mu_R(x, y) = \text{Min} \left\{1, \left[\frac{\mu_B(y)}{\mu_A(x)}\right]\right\} \quad (49)$$

$$\mu_R(x, y) = \text{Max} \{\mu_A(x) \cdot \mu_B(y), [1 - \mu_A(x)]\} \quad (50)$$

$$\mu_R(x, y) = \mu_A(x) \cdot \mu_B(y) \quad (51)$$

$$\mu_R(x, y) = [\mu_B(y)]^{\mu_A(x)} \quad (52)$$

$$\mu_R(x, y) = \begin{cases} \mu_B(y) \text{ for } \mu_B(y) < \mu_A(x) \\ 1 \text{ otherwise} \end{cases} \quad (53)$$

Equations (44) to (53) are valid for all values of $x \in X$ and $y \in Y$. For the case where the input universe of discourse is represented by p discrete elements and the output universe of discourse is represented by q discrete elements, then the fuzzy relation R will be a $p \times q$ matrix.

Equation (44) is known as the classical implication equation or also as *Zadeh's implication*. For $\mu_B(y) < \mu_A(x)$ for all $(x, y) \in (x \bullet y)$, the equation given by (44) reduces to the one given by (45). Equation (46) is called *correlation-minimum* and known as *Mamdani's implication*. This implication may also be found by the fuzzy cross-product of sets A and B, i.e., $R = A \times B$. For $\mu_A(x) \geq 0.5$ and $\mu_B(y) \geq 0.5$ Zadeh's implication reduces to Mamdani's implication. The implication defied by equation (47) is known as *Luckawics implication*. The fuzzy implication relation defined by equation (48) is known as the *bounded sum implication*. Equations (50) and (51) describe two forms of *correlation-product* and are based on the notions of conditioning and reinforcement and also Hebbian type of learning in neuropsychology. This implication method is similar to the one used in artificial neural network computations. Equation (53) is known as *Gödel's implication*. or " α " implication. Equation (50) is suggested by the author and is equally valid for the crisp non-fuzzy cases. The choice of implication equation basically depends on the meaning behind the membership functions defined for the fuzzy sets A and B and also the mechanism by which the (If A THEN B) fact was learned in the first place.

      ii)   *Extension principle*: Let f be a mapping from $x_1 \times x_2 \times \ldots \times x_n$ to the universe y such that $y = f(x_1, x_2, \ldots, x_n)$. The fuzzy extension principle allows us to induce from n fuzzy sets $A_1, A_2, \ldots, A_n$, a fuzzy set B on y through f such that

$$\mu_B(y) = \text{Sup}_{(x_1, x_2, \ldots, x_n) = f^{-1}(y)} \left\{ \text{Min}\left[\mu_{A_1}(x_1), \mu_{A_2}(x_1), \ldots, \mu_{A_r}(x_n)\right] \right\}$$

$$\mu_B(y) = 0 \text{ if } f^{-1}(y) = \emptyset$$

where $f^{-1}(y)$ is the inverse image of y, and $\mu_B(y)$ is the largest among the membership values $\mu_{A_1 \times A_2 \times \ldots \times A_n}(x_1, x_2, \ldots, x_n)$ of the realization of y using n-tuples $(x_1, x_2, \ldots, x_n)$.

For the cases where the system could be described by a closed form function $y = f(x_1, x_2, \ldots, x_n)$, each application of the extension principle to the system will result in a canonical fuzzy rule in the form of

      IF x_1 is A_1 and x_2 is $A_2 \ldots$ and x_n is A_n

      THEN y is B

Therefore, r fuzzy rules may be found by r times application of the extension principle to the system under consideration. Then the implication relations given by equations (44) to (53) can be used to derive the fuzzy relations representing the rules.

There also exist other techniques for obtaining the fuzzy relations which are based on learning and adaptive algorithms. These techniques start from a set of numerical data

Chap. 4   Fuzzy Rule-Based Expert Systems-I   81

derived from observation of the input-output mapping going on in the system. The numerical input-output data is used to come up with linguistic rules or fuzzy relations which correctly fit the available data.

4.3.3   Composition of Fuzzy Relations

Max-Min and Max-Prod methods of composition of fuzzy relation are the two most commonly used techniques for the solution of fuzzy relational equations. In addition, Max-Min, Max-Max, Min-Min, (p, q) composition, Sum-Prod, and Max-Ave techniques are mentioned in literature. Each method of composition of fuzzy relations reflects a special inference machine and has its own significance and applications. Max-Min method is the one used by Zadeh in his approximate reasoning based on linguistic IF-THEN rules. It is claimed that this method of composition of fuzzy relations correctly reflects the approximate and interpolative reasoning used by humans when using natural language propositions for deductive reasoning.

Approximate reasoning involves the following general situation in deductive reasoning:

    IF A THEN B

    IF A^1 THEN ?

The fuzzy relation describing the antecedent $A \to B$ is used to compute the consequent resulting from the application of A^1 as in the following:

    $B = A \cdot R$

    $B^1 = A^1 \cdot R$

    In the following we describe eight different methods for composition of fuzzy relations.

    i)   Max-Min composition

$$y^k = x \cdot R^k$$

$$\mu_{y^k}(y) = \underset{x \in X}{\text{Max}} \left\{ \text{Min}\left[\mu_x(x), \mu_{R^k}(x, y) \right] \right\} \qquad (54)$$

    ii)   Max-Prod composition

$$y^k = x * R^k$$

$$\mu_{y^k}(y) = \underset{x \in X}{\text{Max}} \left[\mu_x(x) \cdot \mu_{R^k}(x, y) \right] \qquad (55)$$

iii) Min-Max composition

$$y^k = x \dagger R^k$$

$$\mu_{y^k}(y) = \underset{x \in X}{\text{Min}} \{\text{Max}[\mu_x(x), \mu_{R^k}(x, y)]\} \quad (56)$$

iv) Max-Max composition

$$y^k = x \circ R^k$$

$$\mu_{y^k}(y) = \underset{x \in X}{\text{Max}} \{\text{Max}[\mu_x(x), \mu_{R^k}(x, y)]\} \quad (57)$$

v) Min-Min composition

$$y^k = x \triangle R^k$$

$$\mu_{y^k}(y) = \underset{x \in X}{\text{Min}} \{\text{Min}[\mu_x(x), \mu_{R^k}(x, y)]\} \quad (58)$$

vi) (p, q) composition

$$y^k = x \cdot^{pq} R^k$$

$$\mu_{y^k}(y) = \underset{x \in X}{\text{Max}_{(p)}} \{\text{Min}_{(q)}[\mu_x(x), \mu_{R^k}(x, y)]\} \quad (59)$$

where

$$\underset{x \in X}{\text{Max}_{(p)}} a(x) = \inf_{(x_1, x_2, \ldots, x_n) \in x} \left\{1, [[a(x_1)]^p + [a(x_2)]^p + \cdots + [a(x_n)]^p]^{\frac{1}{p}}\right\}$$

and

$$\text{Min}_q[a(x), b(x)] = 1 - \min\left\{[1, [1 - a(x)]^q + [1 - b(x)]^q]^{\frac{1}{q}}\right\}$$

vii) Sum-Prod composition

$$y^k = x \times R^k$$

$$\mu_{y^k}(y) = f\left\{\underset{x \in X}{\Sigma}[\mu_x(x) \cdot \mu_{R^k}(x, y)]\right\} \quad (60)$$

where f(•) is a logistic function that limits the value of the function within the interval [0, 1]. This is a composition method commonly used in the artificial neural networks for

Chap. 4 Fuzzy Rule-Based Expert Systems-I   83

mapping between parallel layers in a multi-layer network.

    viii)   Max-Ave composition

$$y^k = x \cdot_{av} R^k$$

$$\mu_{y^k}(y) = \frac{1}{2} \underset{x \in X}{\text{Max}}[\mu_x(x) + \mu_{R^k}(x, y)] \tag{61}$$

4.4 SOLUTION OF A SYSTEM OF FUZZY RELATIONAL EQUATIONS

Once a fuzzy rule-based expert system is put in the canonical form given in Figure 4-4, it is always possible to describe the system under consideration by a system of relational equations, as shown in Figure 4-8. In the following, the general solution of a system of fuzzy relational equations is discussed. The general solution is derived based on two most common techniques for composition of fuzzy relations, i.e., the Max-Min and Max. Prod techniques given in equations (54) and (55).

    i)   Max-Min Method

For a system of disjunctive fuzzy relations, i.e., connected by "or" or "else," the aggregated output y is found based on equations (40) and (55) as in the following:

$$\mu_y(y) = \underset{k}{\text{Max}} \left\{ \underset{x \in X}{\text{Max}} \{\text{Min}[\mu_x(x), \mu_{R^k}(x, y)]\} \right\} \tag{62}$$

where $\mu_y(y)$ is the fuzzy membership function describing the overall output response to the fuzzy input x. For the case of a system with n non-interactive fuzzy inputs, the aggregated output will be in the following form:

$$\mu_y(y) = \underset{k}{\text{Max}} \left\{ \underset{x \in X}{\text{Max}} \{\text{Min}[\mu_{x_1}(x_1), \mu_{x_2}(x_2), \cdots, \mu_{x_n}(x_n), \mu_{R^n}(x_1, x_2, \cdots, x_n, y)]\} \right\} \tag{63}$$

where $x = [x_1, x_2, \ldots, x_n]'$ is the vector of n non-interactive inputs to the fuzzy system.

    For a conjunctive set of fuzzy relations, equations, i.e., connected by "and," equations (35) and (55) are used to find the aggregated output.

    ii)   Max. Prod Method

For a system of conjunctive fuzzy relational equations, the aggregated output y is found based on equations (40) and (56) as illustrated in the following:

$$\mu_y(y) = \underset{k}{\text{Max}} \left\{ \underset{x \in X}{\text{Max}} [\mu_x(x) \cdot \mu_{R^k}(x, y)] \right\} \tag{64}$$

where $\mu_y(y)$ is the fuzzy membership function describing the overall output response to the fuzzy input x.

For the case of a system with n non-interactive fuzzy inputs, the aggregated output will be in the following form:

$$\mu_y(y) = \underset{k}{\text{Max}} \left\{ \underset{x \in X}{\text{Max}} \left[\mu_{x_1}(x_1) \bullet \mu_{x_2}(x_2) \bullet \cdots \bullet \mu_{x_n}(x_n) \bullet \mu_R k(x_1, x_2, \cdots, x_n, y) \right] \right\} \quad (65)$$

and, also, sometimes given in the form:

$$\mu_y(y) = \underset{k}{\text{Max}} \left\{ \underset{x \in X}{\text{Max}} \left[\text{Min} \left[\mu_{x_1}(x_1), \mu_{x_2}(x_2), \cdots, \mu_{x_n}(x_n) \right] \bullet \mu_R k(x_1, x_2, \cdots, x_n, y) \right] \right\} \quad (66)$$

For a disjunctive set of fuzzy relations, equations (35) and (56) are used to find the aggregated output.

4.5 CONCLUSION

As this chapter shows, a set of fuzzy restrictions stated as fuzzy conditional propositions can be put into simple canonical rule-sets. Canonical fuzzy rule-based expert systems may also be represented by a system of fuzzy relational equations that are either conjunctively or disjunctively connected. The chapter presented some techniques for the solution of a system of fuzzy relational equations.

REFERENCES

Dubois, D. and Prade (1980a), *Fuzzy Sets and Systems: Theory and Applications*. Academic Press, New York.

Gupta, M.M., R.K. Ragade, and R.R. Yager, eds. (1979), *Advances in Fuzzy Set Theory and Applications*. North-Holland, New York.

Kaufmann, A. and M.M. Gupta (1985), *Introduction to Fuzzy Arithmetic: Theory and Applications*. Van Nostrand Reinhold, New York.

Klir, G.J. and T.A. Folger (1988), *Fuzzy Sets, Uncertainty, and Information*. Prentice Hall, New Jersey.

Mamdani, E.H. and R.R. Gaines, eds. (1981), *Fuzzy Reasoning and Its Applications*. Academic Press, London.

Pedrycs, W. (1883c), "Some applicational aspects of fuzzy relational equations in systems analysis." *Inern J. of General Systems*, 9 125-132.

Sugeno, M. ed. (1985a), *Industrial Application of Fuzzy Control*. North-Holland, New York.

Zadeh, L.A. (1973), "Outline of a new approach to the analysis of complex systems and decision processes." IEEE *Trans. on Systems, Man, and Cybernetics*, SMC-1. pp. 28-44.

5

FUZZY RULE-BASED EXPERT SYSTEMS II

Nader Vadiee
University of New Mexico

This chapter originally appeared in the journal *Intelligent and Fuzzy Systems: Applications in Engineering and Technology*, March 1993. Used here by permission.

Fuzzy rule-based expert systems (FRBES) models, described by a set of fuzzy conditional statements in the canonical form of (If x is A^k Then y^k is B^k for $k = 1, 2, ..., r$) or a system of conjunctive or disjunctive fuzzy relational equations in the form of ($y^k = x \cdot R^k$ for $k = 1, 2, ..., r$) were discussed in Chapter 4. In Chapter 5, some interesting special cases of FRBES models of dynamic systems and a number of commonly used formats for FRBES models and their solutions are described in details. The fundamental computational processes of FRBES models such as partitioning of input and output spaces into fuzzy partitions and assignment of membership functions to them, fuzzification, inferencing, and defuzzification processes are illustrated. A number of interesting graphical techniques for the above mentioned computational processes are also given. Finally, Fuzzy Associative Memories (FAM) tables are introduced.

5.1 INTRODUCTION

There exists a wide class of complex dynamic processes where the knowledge regarding the functional relationship between the input and output variables may be set up based on numerical and/or non-numerical information. The numerical information data are usually from a limited number of points and the non-numerical information is usually in the form of vague natural language protocols gathered from interviews with the human experts familiar

Chap. 5   Fuzzy Rule-Based Expert Systems-II   87

vague natural language protocols gathered from interviews with the human experts familiar with the input-output behavior or the real time control of the system or process. Complexity in the system models arises as a result of many factors such as (1) high dimensionality, (2) too many interacting variables, and (3) unmodeled dynamics such as nonlinearities, time variations, external noise or disturbance, and system perturbations. Therefore the information gathered on the system behavior, as mentioned above, is never complete, sharp and comprehensive. The investigator needs some interpolative and approximate reasoning methods in order to come up with acceptable and good enough models of the system to the best of his or her available knowledge. No piece of knowledge, be it intuitive or linguistic or whatever other forms, should be discarded on the grounds of the lack of techniques for incorporating that piece of knowledge in the overall decision making process. By the way, this is exactly the way a human expert performs when using all the available clues, intuitions, hunches, learned skills, linguistically stated rules, and formal models to cope with the identification and the real time control of a complex real world system or process. Fuzzy mathematics, which comprises of fuzzy set theory, fuzzy logic, and fuzzy measures has provided a range of mathematical tools that help the investigator to formalize the ill-defined descriptions in the form of linguistically stated IF...THEN... rules into mathematical equations implementable on the conventional digital computers. The artificial neural networks methods, on the other hand, have been used to extract the knowledge present in the learned motor skills of a human operator of the system or a set of numerical measurement results on the system input -output behavior. Integration of these two paradigms along with the proven conventional model-based system identification and control techniques seems to be the new alternative direction to be followed in future intelligent control engineering.

    At the expense of relaxing some of the demands on the precision, a great deal of simplification, ease of computation, speed and efficiency are gained. The ill-defined systems are usually described by fuzzy relational equations. These are expressed in the form of sup-min., max.-prod, or inf.-max. fuzzy compositions. These operations are carried out on classes of membership functions defined on a number of overlapping partitions of the space of possible inputs, possible mapping restrictions, and possible consequent output responses.

    These membership functions are enormously subjective and context dependent. Based on the concept of prototype categorization, degree of similarity, and similarity as a metric distance between an object and a prototype, borrowed from cognitive psychology, three different contexts that can be used for the assignment of membership functions in the area of fuzzy system identification and fuzzy control systems are discussed in this chapter. The general form of a fuzzy relational equation describing a state-feedback controller for a number of systems are derived. The input variables, output variables, and the functional relation describing the controller might be given both in crisp or fuzzy form. In addition, the input-output functional relationship might also be given in the form of a look-up table with a finite number of entries.

    There are some fundamental assumptions that reduce a fuzzy relational equation to a simple canonical form. In this form the input variables are assumed to be non-interactive and the membership function of the points in the state space of the system are assigned based on the degree of similarity of the corresponding output to a prototype output point. The membership functions are further assumed to be linearly dependent on the net distance to a prototype input point. In this way the membership functions will be in one of the rectangular, triangular, or trapezoidal forms. Appropriate nonlinear transformations and/or sensory integration and fusion on input and/or output spaces are often used to reduce a complex fuzzy controller to the canonical form of a fuzzy system model. The net effect of these pre-processings on the input data will be in decoupling and linearizations of system dynamics. Canonical fuzzy systems are also represented by simple fuzzy associative memories (FAMs) which will be introduced later.

Advanced fuzzy controllers use adaptation capabilities to tune the membership functions' vertices or supports or to add or delete rules to optimize the performance and compensate the effects of any internal or external perturbations. Learning fuzzy systems try to learn the membership functions or the rules. Principles of genetics algorithms have been used to find the best string representing an optimal class of input or output symmetrical triangular membership functions.

New generations of fuzzy logic controllers are based on the integration of the conventional and fuzzy controllers. Fuzzy clustering techniques have also been used to extract the linguistic IF...THEN... from the numerical data. In general, the trend is toward the compilation and fusion of different forms of knowledge representation for the best possible identification and control of ill-defined complex systems. The two new paradigms of artificial neural networks and fuzzy systems try to understand a real world system starting from the very fundamental sources of knowledge, i.e., patient and careful observations, measurements, experience, and intuitive reasoning and judgments rather than starting from a preconceived theory or mathematical model.

5.2 SPECIAL CASES OF FUZZY DYNAMIC SYSTEM MODELS

In the following, the fuzzy relational equations describing a number of dynamic systems are illustrated.

i) Consider a discrete first order system with input u and described in a state-space form given by equation (16) of Chapter 4. The basic fuzzy model of such a system has the following form:

$$x_{k+1} = x_k \circ u_k \circ R, \text{ for } k = 1, 2, \ldots \qquad (1)$$

where R is the fuzzy system transfer relation as defined in Chapter 4.

ii) A discrete pth order system with single input u and described in state-space form is given by the fuzzy system equation as follows:

$$x_{k+p} = x_k \circ x_{k+1} \circ \ldots \circ x_{k+p-1} \circ u_{k+p-1} \circ R, \text{ for } k = 1, 2, \ldots \qquad (2)$$

$$y_{k+p} = x_{k+p} \qquad (3)$$

Chap. 5   Fuzzy Rule-Based Expert Systems-II

where y_{k+p} is the single output of the system.

iii) A second order system with full state feedback is described as:

$$u_k = x_k \circ x_{k-1} \circ R \text{ for } k = 1, 2, \ldots \quad (4)$$

$$y_k = x_k \quad (5)$$

iv) A discrete pth order single-input single-output system with full state feedback is represented by the following fuzzy relational equation:

$$u_{k+p} = y_k \circ y_{k+1} \circ \ldots \circ y_{k+p-1} \circ R \text{ for } k = 1, 2, \ldots \quad (6)$$

5.3   SPECIAL CLASSES OF FUZZY RULE-BASED EXPERT SYSTEM MODELS

In the following we will consider five classes of fuzzy rule-based expert system models and their corresponding solutions. The fuzzy systems discussed are described by a collection of r restrictions in the form of linguistic IF...THEN... rules. The systems, described in the following, are single input and single output systems but the results could be easily extended to the p-input and m-output systems where the input x is a p-dimensional vector and output y is an m-dimensional vector.

i) In the first class of FRBES models, the input as well as the output restrictions are given in the form of singletons, i.e.,

```
R¹:   IF  x = x₁    THEN y = y₁        ELSE

R²:   IF  x = x₂    THEN y = y₂        ELSE

  .
  .
  .

Rʳ:   IF x = xᵣ    THEN y = yᵣ .
```

Figure 5-1. A crisp look-up table type of system description.

This is a look-up table type of system description. The value of the output for a given value of input, e.g., $x = x^-$ is equal to the THEN part value of the rule Ri whose IF part matches exactly with the value of input given, i.e.,

$$\text{If } x^- = x_i \text{ -------> } y=y^-=y_i \text{ for } i = 1, 2, ..., r \qquad (7)$$

ii) In the second class of FRBES models, the input restrictions are in the form of crisp sets and the output restrictions are given by singletons, i.e.,

---

R^1 :   IF $x_0 < x < x_1$ THEN $y = y_1$ ELSE

R^2 :   IF $x_1 < x < x_2$ THEN $y = y_2$ ELSE

.
.
.

R^r :   IF $x_{r-1} < x < x_r$ THEN $y = y_r$

Figure 5-2. A crisp look-up table type of system description.

This is a look-up table type of system description. The value of the output for a given value of input, e.g., $x = x^-$ is equal to the THEN part value of the rule R^i whose IF part matches precisely with the value of the input value given, i.e.,

$$\text{If } x_{i-1} < x^- < x_i \text{ -------> } y = y^- = y_i \text{ for } i = 1, 2, ..., r \qquad (8)$$

This is, practically, a piecewise constant approximation of a nonlinear function. The restrictions might also be given as in the following form given in Figure 5-3.

R^1 :   IF $x_0 < x < x_1$ THEN $y = f_1(x)$ ELSE

R^2 :   IF $x_1 < x < x_2$ THEN $y = f_2(x)$ ELSE

.
.
.

R^r :   IF $x_{r-1} < x < x_r$ THEN $y = f_r(x)$.

Figure 5-3. System description using spline functions approximation.

Chap. 5 Fuzzy Rule-Based Expert Systems-II

This is a general spline approximation of a nonlinear function, where $f_1(x)$, $f_2(x)$, ...,$f_r(x)$ are, in general, nonlinear spline functions. The value of the output corresponding to $x = x^-$ is found as in the following:

$$\text{If } x_{i-1} < x^- < x_i \text{ -------> } y = y^- = f_i(x^-) \text{ for } i=1,,...,r \tag{9}$$

iii) In the third class of FRBES models, the input conditions are crisp sets and the output is expressed as a fuzzy set or described by a fuzzy relation. If the THEN part of the rules are given by fuzzy sets defined on the output universe of discourse, the rules will be in the form given below:

R^1 : IF $x_0 < x < x_1$ THEN y is B^1 ELSE

R^2 : IF $x_1 < x < x_2$ THEN y is B^2 ELSE

.
.
.

R^r : IF $x_{r-1} < x < x_r$ THEN y is B^r .

Figure 5-4. FRBES model with crisp IF part conditions.

where B^i for i = 1, 2, ..., r are fuzzy sets defined on the output space. The output of the system corresponding to a given input, e.g., $x = x^-$ is found by matching the given input to one of the IF conditions of the rules, i.e.,

$$\text{If } x_{i-1} < x^- < x_i \text{ -------> } y = y^- = B^i \text{ for } i = 1, 2,...,r \tag{10}$$

The output, in general, is a fuzzy set. A crisp value for the output could be find by a defuzzification process, as will be described in section 5.5, i.e.,

$$y = y^- = DEFUZZ(B^i) \tag{11}$$

If the THEN part restrictions are given in the form of fuzzy relations R^k for k = 1, 2, ..., r, the FRBES model will be in the form given in Figure 5-5.

If the given value of input is $x = x^-$ and $x_{i-1} < x^- < x_i$, then the corresponding value of the output is found using equation (62) or (64) given in Chapter 4, i.e.,

$$\mu_Y(y) = \underset{x=\bar{x}}{Max} \{ Min[1, \mu_{R^i}(\bar{x}, y)]\} \tag{12}$$

$$= \mu_{R^i}(\bar{x}, y)$$

R^1 : IF $x_0 < x < x_1$ THEN restriction R^1 ELSE

R^2 IF $x_1 < x < x_2$ THEN restriction R^2 ELSE
.
.
.
R^r : IF $x_{r-1} < x < x_r$ THEN restriction R^r.

Figure 5-5. FRBES model with fuzzy relations given as its THEN-part restrictions.

where $\mu(\bar{x}) = 1$, i.e., a crisp input is substituted in the equation (62) of Chapter 4. The fuzzy relation R^i is the THEN-part restriction of the ith rule whose IF part condition matches the given input value. A crisp value for the output is found using some defuzzification technique, i.e.,

$$y = \bar{y} = DEFUZZ(Y) = DEFUZZ(R^i) \tag{13}$$

If the Max.-product inference method is used (equation (64) of Chapter 4) the same result will be obtained, i. e.,

$$\mu_Y(y) = \underset{x=\bar{x}}{Max} [1 \cdot \mu_{R^i}(\bar{x}, y)] = \mu_{R^i}(\bar{x}, y) \tag{14}$$

iv) In the fourth class of special FRBES models, the input conditions are given in the form of fuzzy sets defined on input universe of discourse and the output or consequents are given in the form of singletons or generally nonlinear crisp functions. In the former case, the FRBES model is described as shown in Figure 5-6:

Chap. 5   Fuzzy Rule-Based Expert Systems-II

R^1:   IF x is A^1 THEN y = y_1 ELSE

R^2:   IF x is A^2 THEN y = y_2 ELSE

.
.
.

R^r: IF x is A^r THEN y = y_r.

Figure 5-6. FRBES model with a singleton as its THEN-part restriction.

where A^k for k = 1, 2, ..., r are the fuzzy portions defined on the input space X. The output corresponding to a given value of input, e.g., x = x^- is found using one of the following methods:

If $\text{Max}_k [\mu_{A^k}(x^-)] = \mu_{A^i}(x^-)$

then,

$$\bar{y} = y_i \qquad (15)$$

or,

$$\bar{y} = [\sum_{i=1}^{r} \mu_{A^i}(x^-) \cdot y_i] / [\sum_{i=1}^{r} \mu_{A^i}(x^-)] \qquad (16)$$

$$= \sum_{i=1}^{r} w_i \cdot y_i$$

where,

$$w_i = \mu_{A^i}(x^-) / [\sum_{i=1}^{r} \mu_{A^i}(x^-)] \qquad (17)$$

In the second case of this general class, the restrictions are expressed as illustrated in Figure 5-7.

R^1 : IF x is A^1 THEN y = $f_1(x)$ ELSE

R^2 : IF x is A^2 THEN y = $f_2(x)$ ELSE
.
.
.
R^r : IF x is A^r THEN y = $f_r(x)$.

Figure 5-7. FRBES models with spline functions as the THEN-part restrictions.

The output corresponding to a given value of input, e.g., $x = x^-$ is found using one of the following methods:

If $\text{Max}_k [\mu_{A^k}(x^-)] = \mu_{A^i}(x^-)$

then,

$$y^- = f_i(x^-) \qquad (18)$$

or,

$$y^- = [\sum_{i=1}^{r} \mu_{A^i}(x^-) \cdot f_i(x^-)] / [\sum_{i=1}^{r} \mu_{A^i}(x^-)] \qquad (19)$$

$$= \sum_{i=1}^{r} w_i \cdot f_i(x^-)$$

where,

$$w_i = \mu_{A^i}(x^-) / [\sum_{i=1}^{r} \mu_{A^i}(x^-)] \qquad (20)$$

For the case where $f_1(x), f_2(x), ..., $ and $f_r(x)$ are linear functions of the input x, the FRBES model is called a *quasi linear fuzzy model* (QLFM) and for the case where $f_1(x)$, $f_2(x), ..., $ and $f_r(x)$ are generally nonlinear functions of the input x, the FRBES model

Chap. 5 Fuzzy Rule-Based Expert Systems-II

is called a *quasi nonlinear fuzzy model* (QNFM). Figure 5-8 shows a quasi-linear fuzzy system model which is known as static Sugeno's fuzzy model.

R^1: IF x_1 is A_1^1 and x_2 is A_2^1 ... AND x_n is A_n^1
   THEN $y^1 = p_0^1 + p_1^1 \cdot x_1 + p_2^1 \cdot x_2 + \ldots + p_n^1 \cdot x_n$

R^2: IF x_1 is A_1^2 and x_2 is A_2^2 ... AND x_n is A_n^2
   THEN $y^2 = p_0^2 + p_1^2 \cdot x_1 + p_2^2 \cdot x_2 + \ldots + p_n^2 \cdot x_n$.

.
.
.

R^r: IF x_1 is A_1^r and x_2 is A_2^r ... AND x_n is A_n^r
   THEN $y^r = p_0^r + p_1^r \cdot x_1 + p_2^r \cdot x_2 + \ldots + p_n^r \cdot x_n$.

Figure 5-8. Static Sugeno's FRBES model.

By replacing the linear regression model in the static Sugeno's model, given in Figure 5-8, with a difference equation model, we obtain the quasi-linear dynamic fuzzy system model shown in Figure 5-9.

In the FRBES model given in Figure 5-9, u is the input to the system and U is its universe of discourse partitioned into fuzzy partitions $U^1, U^2, \ldots, U^r$ and k represents the kth time instant in a discrete-time system.

v) The most general of FRBES models is the case in which both the input and output restrictions are described by fuzzy sets. The general FRBES model with fuzzy input and fuzzy output is shown in Figure 5-10 and its general solution was discussed in section 4.4 of Chapter 4.

5.4 GRAPHICAL COMPUTATION TECHNIQUES FOR FRBES MODELS

In the following, we consider a two-input and single-output fuzzy system. The results obtained could be easily extended and will hold for fuzzy systems with any number of inputs and outputs. A fuzzy system with two non-interactive inputs x_1 and x_2 and a single output y, is either described by a system of disjunctive or conjunctive relational equations as shown in equation (21):

$$y^k = x_1 \circ x_2 \circ R^k \quad \text{for } k = 1, 2, \ldots, r \qquad (21)$$

R^1: IF u is U^1 x_1 is A_1^1 and x_2 is A_2^1 ... AND x_n is A_n^1

THEN
$$y^1(k) = b_0^1.u(k) + b_1^1.u(k-1) + \ldots b_n^1.u(k-n) \ldots a_1^1.y^1(k-1) - a_2^1.y^1(k-2) - \ldots - a_n^1.y^1(k-n)$$

R^2: IF x_1 is A_1^2 and x_2 is A_2^2 ... AND x_n is A_n^2

THEN
$$y^2(k) = b_0^2.u(k) + b_1^2 u(k-1) + \ldots b_n^2.u(k-n) \ldots - a_1^2.y^2(k-1) - a_2^2.y^2(k-2) - \ldots - a_n^2.y^2(k-n)$$

.
.
.

R^r: IF x_1 is A_1^r and x_2 is A_2^r ... AND x_n is A_n^r

THEN
$$y^r(k) = b_0^r.u(k) + b_1^r.u(k-1 +) \ldots -a_1^r.y^r(k-1) \ldots - a_1^r.y^2(k-1) - a_2^r.y^r(k-2) - \ldots - a_n^r.y^r(k-n).$$

Figure 5-9. FRBES for an *n*-input and single-output system.

R^1: IF x_1 is A_1^1 and x_2 is A_2^1 ... AND x_n is A_n^1
THEN y is B^1

R^2: IF x_1 is A_1^2 and x_2 is A_2^2 ... AND x_n is A_n^2
THEN y is B^2.

.
.
.

R^r: IF x_1 is A_1^r and x_2 is A_2^r ... AND x_n is A_n^r
THEN y is B^r.

Figure 5-10. General form of FRBES for a system with n fuzzy inputs and a single fuzzy output.

Chap. 5 Fuzzy Rule-Based Expert Systems-II

where, for our discussion, "o" stands for a Max.-Min. or Max.-Product composition of fuzzy relations or it may also be, equivalently, described by a collection of linguistic IF...THEN... propositions as in the following:

$$\text{IF } x_1 \text{ is } A_1^k \text{ and } x_2 \text{ is } A_2^k \text{ THEN } y^k \text{ is } B^k \text{ for } k = 1, 2, ..., r \quad (22)$$

In the following, we consider four different cases, i.e., i) the inputs to the system are sharply defined, ii) the inputs to the system are represented by fuzzy sets, iii) input x_1, input x_2 and the system are described in non-fuzzy way, and iv) the inputs x_1 and x_2 are fuzzy sets and the system is characterized by the function $f : X_1 . X_2 \longrightarrow Y$. For the first two cases, we derive the aggregated output based on both Max.-Min. and Max.-product inference techniques.

i) Inputs x_1 and x_2 are sharply defined. The system is described by equation (21), so we will have:

$$\mu(x_1) = \delta(x_1 - \bar{x}_1) = \begin{cases} 1, & \text{for } x = \bar{x}_1 \\ 0, & \text{otherwise} \end{cases} \quad (23)$$

$$\mu(x_2) = \delta(x_2 - \bar{x}_2) = \begin{cases} 1, & \text{for } x = \bar{x}_2 \\ 0, & \text{otherwise} \end{cases} \quad (24)$$

Based on equation (63) of Chapter 4 and for a set of disjunctive rules, the aggregated output will be given by:

$$\mu_Y(y) = \text{Max}_k \{ \text{Max}_{x_1 \in X_1} \{ \text{Max}_{x_2 \in X_2} \{ \text{Min}[\mu_{X_1}(x_1), \mu_{X_2}(x_2), \mu_{R^k}(x_1, x_2, y)] \} \} \} \quad (25)$$

Substituting the values of membership functions given in equations (23) and (24) we get:

$$\mu_Y(y) = \text{Max}_k \{ \text{Max}_{x_1 \in X_1} \{ \text{Max}_{x_2 \in X_2} \{ \text{Min}[1, 1, \mu_{R^k}(x_1, x_2, y)] \} \} \} \quad (26)$$

$$= \text{Max}_k [\mu_{R^k}(\bar{x}_1, \bar{x}_2, y)]$$

Therefore, in order to find the value of the aggregated output at any point y, we have to find the value of all the fuzzy relations R^k for k = 1, 2, ..., r evaluated at $x_1 = \bar{x}_1$, $x_2 = \bar{x}_2$, and y = y and then take the maximum value. For the case where the fuzzy system is described by a set of linguistic IF...THEN... rules as in equation (22), based on Mamdani's implication relation given by the equation (46) of Chapter 4, this will turn out to be given by:

$$\mu_Y(y) = \text{Max}_k \{ \text{Min}[\mu_{A_1^k}(\bar{x}_1), \mu_{A_2^k}(\bar{x}_2), \mu_{B^k}(y)] \} \quad (27)$$

Equation (27) has a very simple graphical interpretation which is illustrated in Figure 5-11. In order to find a sharp a crisp value for the aggregated output, some appropriate defuzzification technique could be employed, i.e.,

$$\bar{y} = \text{DEFUZZ}[\mu_Y(y)]$$

For Max.-product composition technique, the aggregated output is given by equation (65) of Chapter 4, i.e.,

$$\mu_Y(y) = \text{Max}_k \{ \text{Max}_{x_1 \in X_1} \{ \text{Max}_{x_2 \in X_2} \{ [\mu_{X_1}(x_1) . \mu_{X_2}(x_2) . \mu_{R^k}(x_1, x_2, y)] \} \} \}$$

Chap. 5 Fuzzy Rule-Based Expert Systems-II

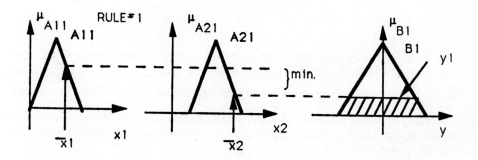

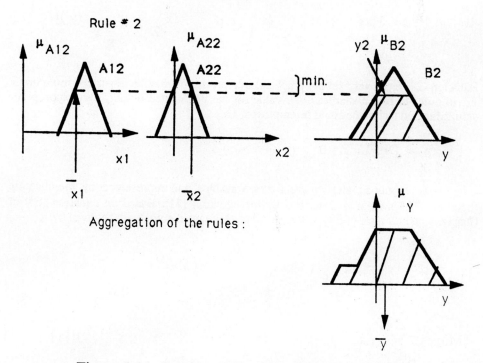

Figure 5-11. Graphical Max.-Min. inference method.

Substituting the values of membership functions given in equations (18) and (19) we get:

$$\mu_Y(y) = \text{Max}_k \{\text{Max}_{x_1 \in X_1} \{\text{Max}_{x_2 \in X_2} \{[\delta(x_1 - \bar{x}_1) \cdot \delta(x_2 - \bar{x}_2)$$

$$\cdot \mu_{R^k}(\bar{x}_1, \bar{x}_2, y)]\}\} = \text{Max}_k [\mu_{R^k}(\bar{x}_1, \bar{x}_2, y)] \quad (28)$$

therefore, in order to find the value of the aggregated output at any point y we have to find the value of all the fuzzy relations R^k for k = 1, 2, ..., r evaluated at $x_1 = \bar{x}_1$, $x_2 = \bar{x}_2$, and y = y and then take the maximum value. For the case where the fuzzy system is described by a set of linguistic IF...THEN... rules as in equation (22), based on correlation product relation given by the equation (51) of Chapter 4, this will turn out to be given by:

$$\mu_Y(y) = \text{Max}_k \{[\mu_{A_1^k}(\bar{x}_1) \cdot \mu_{A_2^k}(\bar{x}_2) \cdot \mu_{B^k}(y)]\} \quad (29)$$

Equation (29) has a very simple graphical interpretation which is illustrated in Figure 5-12. In order to find a sharp a crisp value for the aggregated output, some appropriate defuzzification technique could be employed, i.e.,

$$\bar{y} = \text{DEFUZZ}[\mu_Y(y)] \quad (30)$$

ii) Inputs x_1 and x_2 are fuzzy variables and represented by membership functions. The system is described by the equation (21). Based on equation (63) of Chapter 4 and for a set of disjunctive rules the aggregated output will be given by:

$$\mu_Y(y) = \text{Max}_k \{\text{Max}_{x_1 \in X_1} \{\text{Max}_{x_2 \in X_2}$$

$$\{\text{Min}[\mu_{X_1}(x_1), \mu_{X_2}(x_2), \mu_{R^k}(x_1, x_2, y)]\}\}\} \quad (31)$$

For the case where the fuzzy system is described by a set of linguistic IF...THEN... rules as in equation (22), based on Mamdani's implication relation given by the equation (46) of Chapter 4, this will turn out to be given by:

Chap. 5 Fuzzy Rule-Based Expert Systems-II

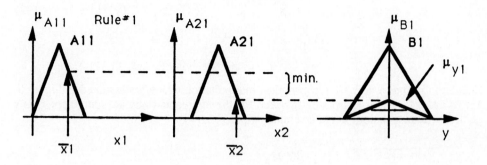

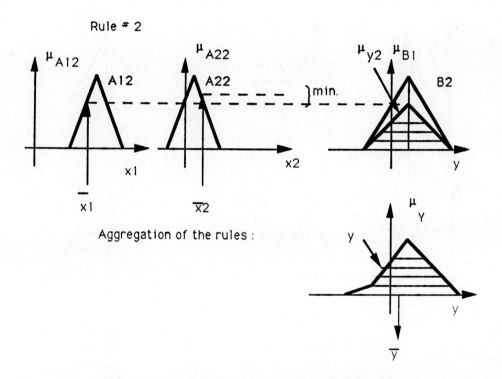

Figure 5-12. Graphical Max.- Product method.

$$\mu_Y(y) = \text{Max}_k \{\text{Max}_{x_1 \varepsilon X_1} \{\text{Max}_{x_2 \varepsilon X_2} \{\text{Min}[\mu_{X_1}(x_1),$$

$$\mu_{X_2}(x_2), \mu_{A_1^k}(x_1), \mu_{A_2^k}(x), \mu_{B^k}(y)]\}\}\} \tag{32}$$

Equation (32) has a very simple graphical interpretation which is illustrated in Figure 5-13. In order to find a sharp a crisp value for the aggregated output, some appropriate defuzzification technique could be employed, i.e.,

$$\bar{y} = \text{DEFUZZ}[\mu_Y(y)] \tag{33}$$

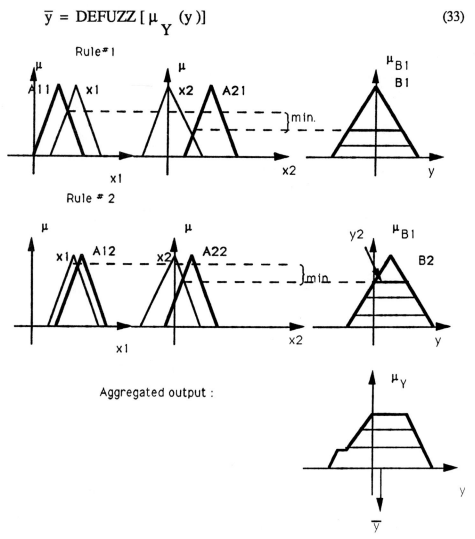

Figure 5-13. Graphical max.-min. method for fuzzy inputs.

Chap. 5 Fuzzy Rule-Based Expert Systems-II

For Max.-product composition technique, the aggregated output is given by the equation (65) of Chapter 4, i.e.,

$$\mu_Y(y) = \underset{k}{\text{Max}} \left\{ \underset{x_1 \in X_1}{\text{Max}} \left\{ \underset{x_2 \in X_2}{\text{Max}} \left\{ [\mu_{X_1}(x_1) \cdot \mu_{X_2}(x_2) \cdot \mu_{R^k}(x_1, x_2, y)] \right\} \right\} \right\} \quad (34)$$

For the case where the fuzzy system is described by a set of linguistic IF...THEN... rules as in equation (17), based on correlation product implication relation given by the equation (51) of Chapter 4, this will turn out to be given by:

$$\mu_Y(y) = \underset{k}{\text{Max}} \left\{ [\mu_{A_1^k}(\bar{x}_1) \cdot \mu_{A_2^k}(\bar{x}_2) \cdot \mu_{B^k}(y)] \right\} \quad (35)$$

In order to find a sharp a crisp value for the aggregated output, some appropriate defuzzification technique could be employed, i.e.,

$$\bar{y} = \text{DEFUZZ}[\mu_Y(y)] \quad (36)$$

iii) For the case where the input x, output y, and the system are described in a non-fuzzy way, based on equations (23) and (24), the output will turn out to be as follows:

$$\mu_Y(y) = \underset{x_1 \in X_1}{\text{Max}} \left\{ \underset{x_2 \in X_2}{\text{Max}} \left\{ \text{Min}[\delta(x_1 - \bar{x}_1), \delta(x_2 - \bar{x}_2), \delta_{y,(\bar{x}_1, \bar{x}_2)}] \right\} \right\} \quad (37)$$

where,

104 Chap. 5 Fuzzy Rule-Based Expert Systems-II

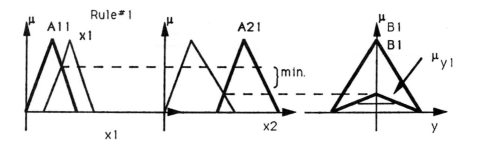

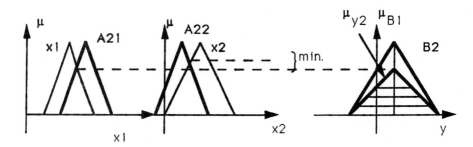

Aggregation of the rules:

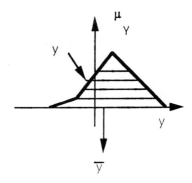

Figure 5-14. Graphical max.-product method for fuzzy inputs.

$$\delta_{y, (\bar{x}_1, \bar{x}_2)} = \begin{cases} 1, & \text{for } (\bar{x}_1, \bar{x}_2) = f^{-1}(y) \\ 0, & \text{otherwise} \end{cases} \quad (38)$$

Note that the system is described as:

Chap. 5 Fuzzy Rule-Based Expert Systems-II

$$y = f(x_1, x_2)$$

Equation (37) leads to:

$$\mu_Y(y) = \begin{cases} 1, & \text{for } y = f(\bar{x}_1, \bar{x}_2) \\ 0, & \text{otherwise} \end{cases} \quad (39)$$

So we get back the crisp form of the system described by $y = f(x_1, x_2)$ and this proves that the conventional crisp mathematics could be considered as a special case of the fuzzy mathematics.

iv) Inputs x_1 and x_2 are fuzzy sets and the system is being described by a crisp function, e.g., $y = f(x_1, x_2)$. The output of the system is found by virtue of Zadeh's Extension Principle, discussed in Chapter 4, i.e.,

$$\mu_Y(y) = \underset{(x_1, x_2) \in f^{-1}(y)}{\text{Max}} \{ \text{Min}[\mu_{X_1}(x_1), \mu_{X_2}(x_2)] \} \quad (40)$$

The output is generally a fuzzy set which could be defuzzified if desired, i.e.,

$$y = \text{DEFUZZ}[(\mu_Y(y))] \quad (41)$$

5.5   DEFUZZIFICATION PROCESS

Defuzzification process is defined as the conversion of a fuzzy quantity, represented by a membership function, to precise or crisp quantity. In the following, three commonly used techniques for defuzzification of fuzzy quantities are described.

i) Max. method- In cases where the membership function, characterizing the fuzzy quantity, has a unique peak point the crisp value corresponding to the peak of the function is taken to be the best representative value of the fuzzy quantity, i.e.,

$$\bar{y} = \text{DEFUZZ}[\mu_Y(y)]$$

where,

$$\underset{y \in Y}{\text{Max}}[\mu_Y(y)] = \mu_Y(\bar{y}) \quad (42)$$

Figure 5-15 illustrates the Maximum defuzzification method.

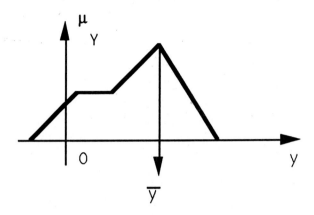

Figure 5-15. Maximum membership method of defuzzification.

ii) Centroid method- In this method of defuzzification, the weighted average of the membership function or the center of the gravity of the area bounded by the membership function curve is computed to be the most typical crisp value of the fuzzy quantity, i.e.,

$$\bar{y} = \frac{\int \mu_Y(y) \cdot y \, dy}{\int \mu_Y(y) \, dy} \quad (43)$$

This defuzzification method is depicted in Figure 5-16.

iii) Height method- This defuzzification technique is valid only for the case where the output membership function is an aggregated union result of symmetrical functions. Assuming that

$$\mu_Y(y) = \text{Max}[\mu_{y_1}(y), \mu_{y_2}(y), ..., \mu_{y_r}(y)] \quad (44)$$

and,

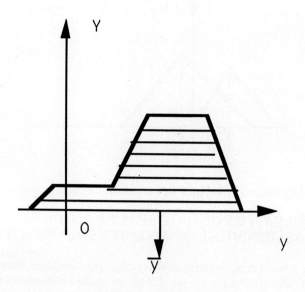

Center of Gravity

Figure 5-16. Centroid method for defuzzification.

$$\mu_{y^k}(\bar{y}^k) = \text{Max}\,[\mu_{y^k}(y)] \qquad (45)$$

then, the defuzzified output is obtained by:

$$\bar{y} = \frac{\sum_{k=1}^{k=r} \mu_{y^k}(\bar{y}^k) \cdot \bar{y}^k}{\sum_{k=1}^{k=r} \mu_{y^k}(\bar{y}^k)} \qquad (46)$$

Figure 5-17 illustrates the height defuzzification method for r=2.

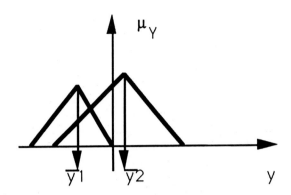

Figure 5-17. Height defuzzification technique.

5.6 PARTITIONING OF SPACES AND ASSIGNMENT OF MEMBERSHIP FUNCTIONS

The fuzzy relation R has a local property an is very context dependent. There are three most intuitive and frequently used ways for the definition of the membership functions of the input conditions as well as the output consequences. These methods are explained based on the prototype categorization, degree of similarity, and "similarity as distance" concepts. It is assumed that the fuzzy system under consideration has n non-interactive inputs and the knowledge regarding the functional relation of the system is extracted from the interviews with an human expert. The expert knowledge can be put in the form of linguistic IF...THEN... rules such as " if the input is medium positive then the output is negative low", or "if the input is low the output is medium", and etc. In this way, the expert is simply giving the functional relation between pairs of prototype points "medium" and "low" in the input space and the prototype points "negative low" and " medium" in the output space. A crisp binary relation between prototype input points and prototype output points is all we are able to extract from the expert knowledge. The exact numerical values of these prototype points and the corresponding membership functions that these prototype points represent is opaque to us. These are highly subjective and intuitive for the human expert. The human expert may not be even able to give the values. Some intuitive methods for assigning membership functions to linguistic values, used by human experts in their propositions about a system, will be described in the following. There exist two different situations.

(1) There are situations where the range of the input and output variables are somehow known. In this case the interval from the lower limit to the higher limit is divided into n equal partitions and the mid points of each partition is taken to represent the prototype point, i.e., the membership equal to unity. For the two extreme partitions the minimum and maximum end points are assumed to be the corresponding prototype points. It is intuitively assumed that the degree of membership of each point in the fuzzy set represented by a prototype point. It is further assumed that the value of membership function corresponding to a typical prototype point is equal to zero at all other prototype points proportional to the distance of that point to the prototype point. Based on this assumption, n triangular shape membership functions will be defined on the interval. This

Chap. 5   Fuzzy Rule-Based Expert Systems-II   **109**

procedure can be used for all the input and output scalar variables. Figure 5-18 is an illustration of this procedure. Some intuitive linguistic values, such as negative big, negative medium, etc., might be given to the fuzzy sets P1, P2, etc. Once the membership functions are found the fuzzy system relations, i.e., R's are found based on the method discussed in chapter 4.

(2)   It is also possible to ask the human expert to give the value(s) of the most typical point, points, or intervals (prototypes) which correctly represent a linguistic category used in his propositions on the system. A similar procedure to the one discussed for case (1), can be used to come up with a class of membership functions for each variable. Figure 5-19 illustrates an example of the second situation. Fuzzy sets P1, P2, etc. take the linguistic values and terms which the expert has used for designating his prototypes or have been used by him or her in the natural language propositions in the form of IF...THEN... rules. Once the membership functions are found this way, the fuzzy relations, i.e., R's are found based on implication relations in Chapter 4.

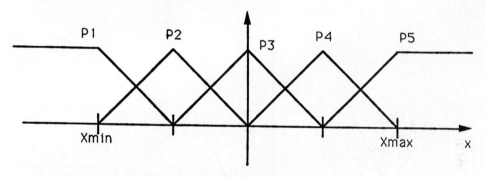

Figure 5-18. Fuzzy partitions for equally spaced prototype points.

5.7   FUZZY ASSOCIATIVE MEMORY (FAM) TABLES

Consider a fuzzy system with n non-interactive inputs and a single output. We assume that each input universe of discourse, i.e., $X_1, X_2, ..., X_n$ is partitioned into k fuzzy partitions. Based on the canonical FRBES model given for such a system in Chapter 4 equation, the total number of possible rules is given by:

$$l = (k+1)^n \qquad (47)$$

where l is the maximum possible number of canonical rules. The number of rules, r, necessary to describe a fuzzy system is much less than l, i.e., $r << l = (k+1)^n$. This is due to the interpolative reasoning capability of the fuzzy inference engine.

For a small number of inputs, e.g., n = 1 or n = 2, or n = 3, there exist a compact form of representing the FRBES model. These forms are illustrated for n = 1 in Figure 5-20. The compact form is called a fuzzy associative memory table or FAM table. As it is understood from the FAM table, the FRBES model is actually representing a generally nonlinear mapping from the input space of the fuzzy system to the output space of the

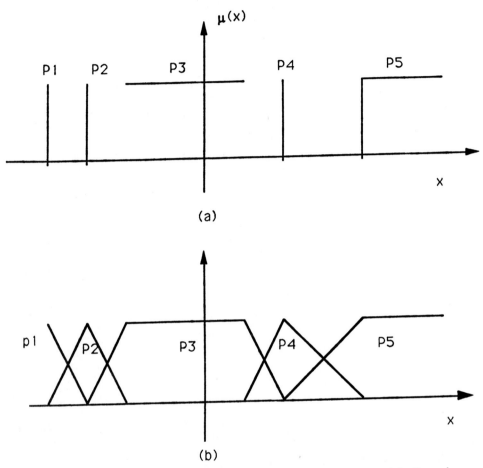

Figure 5-19. (a) Prototype points and intervals, (b) membership functions.

fuzzy system. In this mapping, the patches of the input space are being applied to the patches in the output space. Each rule or, equivalently, each fuzzy relation from input to the output, is actually representing a fuzzy point of data that characterizes the nonlinear mapping from input to the output.

5.8   CONCLUSIONS

In Chapter 5, some interesting special cases of FRBES models of dynamic systems and a number of commonly used formats of FRBES models and their solutions were described in details. The fundamental computational processes of FRBES models, such as, fuzzification, inferencing, input and output spaces partitioning and assignment of membership functions, and defuzzification processes were illustrated. A number of

Chap. 5 Fuzzy Rule-Based Expert Systems-II

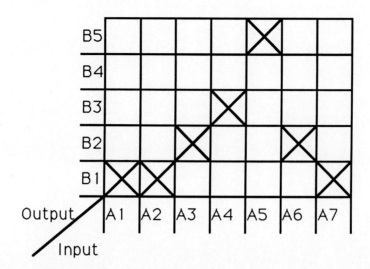

Figure 5-20. FAM table for a single-input single-output FRBES model.

interesting graphical techniques for the above mentioned computational processes were also given. Finally, Fuzzy Associative Memories (FAM) tables were introduced.

REFERENCES

Dubois, D. and Prade (1980a), *Fuzzy Sets and Systems: Theory and Applications.* Academic Press, New York.

Gupta, M.M., R.K. Ragade, and R.R. Yager, eds. (1979), *Advances in Fuzzy Set Theory and Applications.* North-Holland, New York.

Mamdani, E.H. and R.R. Gaines, eds. (1981), *Fuzzy Reasoning and Its Applications.* Academic Press, London.

Sugeno, M. ed. (1985a), *Industrial Application of Fuzzy Control.* North-Holland, New York.

6

FUZZY LOGIC SOFTWARE AND HARDWARE

Mo Jamshidi
University of New Mexico

In the last 5 years many fuzzy logic software and hardware products have begun to appear in the market in the USA, Japan, and other countries. The object of this chapter of the book is to overview some of the available software and hardware. Simulation and real-time fuzzy control examples are provided using some of the software and hardware products discussed. The reader should note that it is always difficult to do a fair and objective job in any overview of this nature; one good reason for this is that not all of the software or hardware were available to the author and his associates.

6.1 FUZZY LOGIC SOFTWARE

A number of fuzzy logic software programs are available in the market or at private establishments. These fuzzy logic software programs are summarized in Table 6-1. Perhaps the most common among these from the point of view of fuzzy control design is Togai InfraLogic's Fuzzy-C Expert System [1]. The next section gives a brief introduction into Fuzzy-C.

Chap. 6   Fuzzy Logic Software and Hardware   113

6.1.1 Togai's Fuzzy-C Expert System

This software is an environment for writing, testing, debugging, and using fuzzy logic expert systems. It does NOT allow the user to perform a complete fuzzy logic control simulation by itself. It does allow the user to write knowledge bases (rule bases) in a high level language called *TIL Fuzzy Programming Language*; it then generates portable ANSI or Kernighan and Richie C source codes for the subroutines and data necessary to implement a specific expert system. A typical .TIL file written in this language consists of the following segments :

1. Input/output identification
2. Variable data types
3. Membership functions definitions
4. Fuzzy rules definitions
5. Input/output and fuzzy rules connections

Program	Computer Media	Developer(s)	Special features
1. Fuzzy-C	Personal Computers and Macintosh	Togai InfraLogic Irvine, CA	Fuzzy TIL Language
2. TIL Shell	Personal Computer	Togai InfraLogic Irvine, CA	Graphical design under MS Windows
3. Fuzzy Micro Controller	Personal Computers	NeuraLogix Sanford, FL	Runs with PC hardware card for real-time control
4. FLCG	Macintosh	Univ. New Mexico Albuquerque, NM	Fuzzy logic code generator
5. FULDEK	Personal Computers	Bell Helicopter Textron, Inc. Fort Worth, TX	Complete fuzzy control simulation environment under MS Windows
6. FL_Control	Personal Computers	Texas A&M U. College Stn., TX	Adaptive and non-adaptive fuzzy controllers
7. FIDE	Personal Computers	Aptronix, Inc. Palo Alto, CA	Software, Simulation Code Generation

Table 6-1. A partial listing of fuzzy logic and control software.

As an example, assume that a given system is to be designed using a fuzzy control of PD type with the following characteristics: Controller inputs are error e, change in error is DError, and the controller output is the armature voltage of a DC motor u. Assume that the error, change in error, and armature voltage have membership functions shown in Figure 6-1. Moreover assume that two of the fuzzy rules are given below: (1) "if e is positive medium and DError is zero then armature voltage is negative medium," (2) "If error is negative small and DError is positive small then armature voltage is zero."

The second listing for variable error e is shown in Table 6-3. The appropriate rule base for this test example is shown in Table 6-4, while Table 6-5 shows the a piece of .TIL code to connect the input variables to the rule base and the rule base to the output variables.

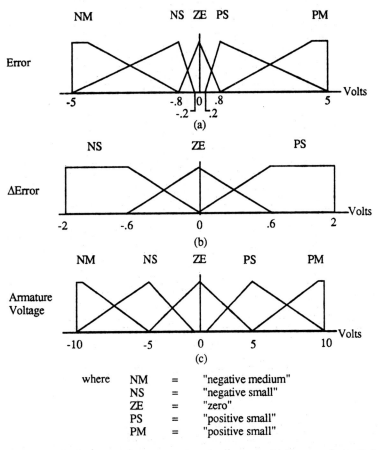

Figure 6-1. Membership functions a) Input error, b) Input change in error, c) Output armature voltage.

It is noted that the file "TEST.TIL" consists of all the above four listings, i.e., variable definitions, membership function, fuzzy rule set, and the connect statements. In

Chap. 6   Fuzzy Logic Software and Hardware   **115**

addition, a driver program must be written in C to accommodate the .TIL file. One such file is shown in Table 6-6.

Table 6-2 shows a partial listing of a file named "TEST.TIL" for this test problem. The software formulation in Table 6-2 is discussed later in the "inverted pendulum" example.

```
           /* Test program in .TIL code to illustrate LISTINGS of TIL Fuzzy_C System
*/
           PROJECT TEST

           /* LISTING -1 Fuzzy_C systems */
           VAR Error
                TYPE signed byte        /* C type of "signed byte" */
                MIN -128                /* minimum of universe of discourse */
                MAX 127                 /* maximum of universe of discourse */
           VAR DError
                ....
           /* Fuzzy Variable Membership Functions */

           * Membership functions for DError (ZE, PS, NS). */

      MEMBER ZE
           POINTS -30 0 0 1 30 0
      END

      MEMBER PS
           POINTS 0 0 30 1 100 1
      END

      MEMBER NS
           POINTS -100 1 -30 1 0 0
      END
           END /* end of DError definition */

           /* End of LISTING -1 Fuzzy-C System */
```

Table 6-2. TIL listing variable DError and its membership functions using Togai's Fuzzy-C system.

6.1.2 Bell Helicopter's Fuzzy Logic Development Kit - FULDEK

FULDEK is a fuzzy logic development system developed by Dreier [2] at Bell Helicopter Textron, Inc. It is a Microsoft Windows application program written in Microsoft Visual Basic language. Operating the program requires MS-DOS, WINDOWS, and a mouse installed on the computer. The program has several files called FORMS (*.FRM), one executable, and a dynamic link library. All of these should be present on the drive and

116 Chap. 6 Fuzzy Logic Software and Hardware

directory from which you plan to execute. To start the program: at the MS-DOS prompt, type: >WIN FLDK2

When all forms are loaded, the EDITOR OPTION form will appear and the user is on its way.

FULDEK has two main forms, the EDITOR OPTION and the RUN OPTION. Each form has a menu bar at the top with items that can be actuated either by clicking with the mouse or by pressing the ALT key in combination with the underlined letter of the menu item. Some menu items have drop-down menus of their own, and some of these have further sub-menus. The presence of sub-menus is indicated with a filled triangle beside the drop-down menu item.

Editor Option

To modify an existing name, "click" it in the Known Variable List. This will display the values of the properties in the yellow box to the right. In Figure 6-2, the variable TORQUE has been selected. The user can see at once that the TORQUE variable is an

```
/*  Test program in .TIL code to illustrate LISTINGS of TIL Fuzzy-C System */
        PROJECT TEST
        /* LISTING -1 Fuzzy-C systems */
        VAR Error
                TYPE signed byte        /* C type of "signed byte" */
                MIN -128                /* minimum of universe of discourse */
                MAX 127                 /* maximum of universe of discourse */
        VAR DError
                ....
        /* Fuzzy Variable Membership Functions */

        * Membership functions for DError (ZE, PS, NS). */
        MEMBER ZE
            POINTS -30 0 0 1 30 0
        END

        MEMBER PS
            POINTS 0 0 30 1 100 1
        END

        MEMBER NS
            POINTS -100 1 -30 1 0 0
        END

                END /* end of DError definition */

            /* End of LISTING -1 Fuzzy-C System */
```

Table 6-3. TIL listing for variable error and its membership functions.

Chap. 6 Fuzzy Logic Software and Hardware

```
/* Fuzzy IF-THEN Rules Set */

/* Rules for response */

FUZZY Alignment_rules
    RULE Rule 1
        IF Error IS PM AND DError IS ZE THEN
            Speed IS NM
    END

    RULE Rule 6
        IF Error IS NS AND DError IS PS THEN
            Speed IS ZE
    END

/* End of Listing - 2  */
```

Table 6-4. TIL listing rule base for "test" problem.

```
/* The following three CONNECT Objects specify that Error
 * and DError are inputs to the Alignment_rules knowledge base
 * and Speed output from Alignment_rules.
 */

    CONNECT
        FROM Error
        TO Alignment_rules
    END

    CONNECT
        FROM Alignment_rules
        TO Speed
    END

END

/* End of Listing -3  */
```

Table 6-5. CONNECT code for "test" problem on Fuzzy-C system.

output variable, is evaluated proportionally (that is, immediately), and has a scale factor of 60 ft-lbs. This variable could be removed from the known list by selecting **Remove**. To modify the values, the user simply "clicks" the option buttons or "clicks" the number fields and enters the value desired.

```
Driver Program C Code

/*
    Fuzzy Laser Beam Tracker driver program
*/

#include "address.h"
#include "stdio.h"

main()

fp = fopen(datafile, "w");

fprintf(fp, "FUZZY BEAM ALIGNMENT DATA\n\n");

fprintf(fp, "X-ERROR    Y-ERROR\n");

for (i = 0; i <= 499; i++)
{
    fprintf(fp, "%f    %f\n," x_data[i], y_data[i]);
}

fclose(fp);

printf("\n\nFile Transfer Complete.\n\n"); }
```

Table 6-6. Drive code (Main Program) to go with TEST.TIL file in Fuzzy-C system.

To add a variable, "click" **New** and enter the variable name. There are four properties which must be defined: the Type property, the Evaluation property, the Initial Condition property and the Scale Factor property. Each of these is discussed below. To select a Type, simply "click" the desired type from the given list. **Input** means this variable is purely an antecedent variable. **Output** means the variable is only a conclusion variable. **Feedback** means the variable can be both an input and output. This means production rules of the form:

IF X IS A THEN Y IS B

IF Y IS B THEN Z IS C

can be evaluated. Note, however, that the second rule does not receive the value of B from the first rule until the next pass through the rule base. **Constant** means a conclusion variable is given a constant value if the antecedent is true to any non-zero degree. For instance, if Y is declared a constant type and the degree of membership of X in A = $m_A(x)$ = 0.1, then rule:

IF X IS A THEN Y IS B

gives the function Y the fuzzy value B with membership degree 1.0.

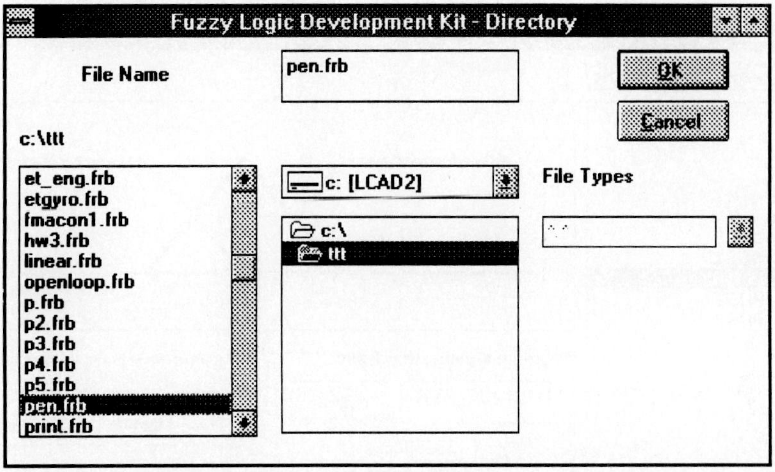

Chap. 6   Fuzzy Logic Software and Hardware   119

Figure 6-2. FULDEK's Directory window.

Memberships

This sub-command lets the user edit existing membership functions or add new functions. An example of this panel is shown in Figure 6-3. All functions have four vertices at the points (X1,0), (X2,1), (X3,1), and (X4,0), thus only the x values are required. These points define a trapezoid, though a triangle may be created if X2=X3. There is no restriction on the X values, except that the functions must be defined such that X1 ≠ X2 ≠ X3 ≠ X4. Thus

$$X1 = -1.5$$
$$X2 = -1.0$$
$$X3 = -1.0$$
$$X4 = -0.5$$

is a valid triangle centered at -1.0, but

$$X1 = -1.0$$
$$X2 = -0.5$$
$$X3 = -0.8$$
$$X4 = -0.2$$

is illegal since X3 = X2.

Compose

When this option is selected, the form shown in Figure 6-4 appears. All known "IF" and "THEN" variables are listed, as are all known membership functions. To build a rule, follow the instructions in the yellow instruction field. In general, you will follow this path:

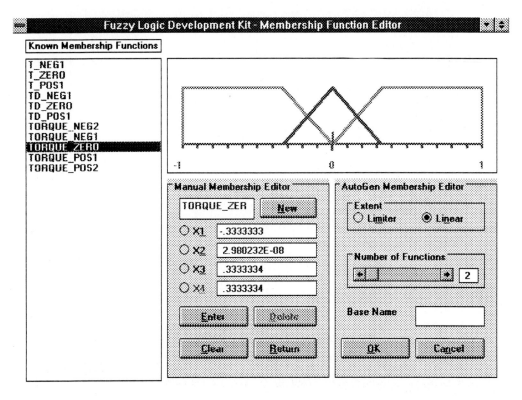

Figure 6-3. FULDEK's Membership Directory window.

1. Select an IF variable
2. Select IS or IS NOT
3. Select a membership function
4. Select AND, OR, or THEN
4a. Select AND, then go to Step 1
4b. Select OR, then go to Step 1
4c. Select THEN, then go to Step 5
5. Select a THEN variable
6. Select IS only
7. Select a membership function
8. Select END

At this point, the composed rule is displayed in the dialog box to the left. **Accept** it and the rule goes into the Fuzzy Rule Base; **Reject** it, and the rule is cast aside. In either event, the user is returned to the beginning of the cycle. From here, the user can write another rule; search for a rule using **Next**, **Previous**, or **Find**; **Delete** a rule; or simply **Return**.

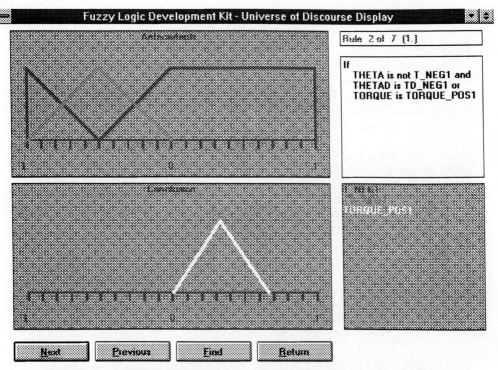

Figure 6-4. FULDEK's Universe of Discourse Display window.

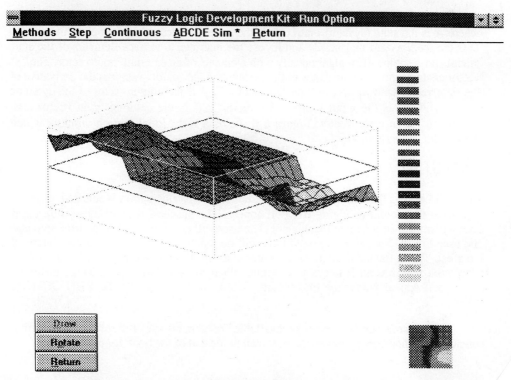

Figure 6-5. FULDEK's Run Option window showing
a 3-D (control surface) plot.

Run Option

If a 3-D plot is to be made, the **D**raw command will produce a surface point, an example of which can be seen in Figure 6-5. The **R**otate command will rotate the surface map 90 degrees counter-clockwise for a different look at the surface. This command can be used as many times as is desired.

If a God's Eye View plot is made, then a map (Fuzzy Associate Memory, or FAM) can be obtained. Note that unlike the 3-D graphics, the Rotate command is not activated for this option. If an X-Y plot is made, the input variable is swept along the abscissa and the output variable is plotted as the ordinate. Again, the Rotate command is not activated for this option. **R**eturn sends you back to the previous page, and "clicking" the **R**eturn button will return you to the RUN OPTION form.

ABCDE Sim

This sub-command allows you to link your fuzzy rule base to a dynamic linear model represented in state space with 5 matrices. The matrices will be discussed more in the next chapter, but for the sake of clarity, the basic state space model reads:

$$
\begin{aligned}
dx/dt &= A\,x + B\,u \\
y_u &= C_u\,x + D_u\,u \\
e &= C_e\,x + E_e\,u_e
\end{aligned}
$$

where A is the state (system) matrix, B is the control matrix, C is the output matrix and D is the feedforward matrix. C_e and E_e are the matrices to extract elements of the state model and combine them algebraically with elements of an external input vector. The "x" vector contains the dynamic states of the system and the "dx/dt" vector is the derivative of the state vector with respect to time. The vector "y_u" is the output vector of the dynamic system and "u" is the input vector. The vector "u_e" is the external input vector and, finally, "e" is an error vector. Figure 6-6 shows a Get Model Option window which defines the 2x2 system matrix A.

RUN **S**IM

RUN **S**IM allows the user to begin the simulation. When this option is selected, a screen similar to that found in Figure 6-7 will appear. The abscissa and ordinates, titles, and time will appear in the large blackboard. The user will be prompted for the time step and end time to use. Sometimes the end time will be reduced if the number of time steps is too great. The initial conditions for the states were specified in the model date file, so the simulation now has all it needs to execute. When the simulation is finished, a Return button will appear below the blackboard. "Click" it to go back to the RUN OPTION screen.

This software will soon be available for general use. Interested readers may contact the author directly or use the post card at the end of the book for more details.

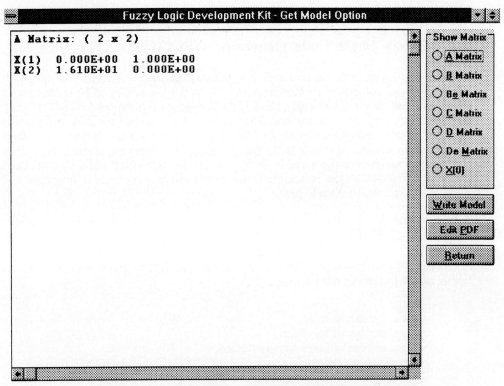

Figure 6-6. FULDEK's Get Model Option window.

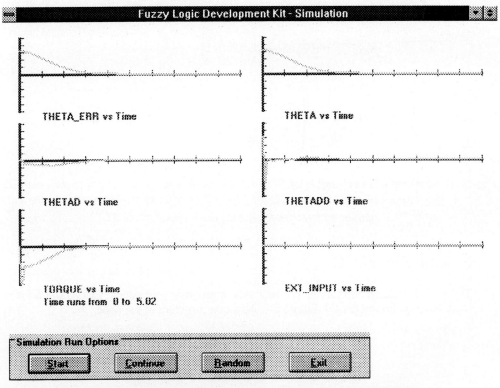

Figure 6-7. FULDEK's System Simulation window.

6.1.3 Fuzzy Logic Code Generator - FLCG

FLCG (Fuzzy Logic Code Generator) is an application package written for the Macintosh line of personal computers by Rashid-Alang [3] at the University of New Mexico. It generates a code in the C language (Think C by Symantec Corporation) for implementing user-specified fuzzy logic applications. The generated code can be "attached" to a process model to simulate fuzzy logic control of the process. This code also writes to a default filename the intermediate results of the fuzzy logic control program, membership values of any input variables in the variables fuzzy subsets, the rules fired and their associated strengths. These results can be used to study the effectiveness or relative importance of any rules in the controller's rule-base.

Features of FLCG

FLCG has been written mostly for educational use and light research-oriented projects. Below are the features of this package :

- Max. number of rules : 40
- Max. number of variables : 7
- Max. number of fuzzy subsets/variable : 7
- Fuzzification operator : Gaussian like (modifiable) - see *Note 1*
- Defuzzification scheme : Simplified centroid method - see *Note 2*
- Fuzzy connectives : AND, OR

Note 1. $m_{\overline{x}}(x) = \exp\left(-\frac{(x-\overline{x})^2}{2s}\right)$.

Note 2. $u = \left(\dfrac{\sum_{i=1}^{n} m_{R_i} c_i}{\sum_{i=1}^{n} m_{R_i}}\right)$,

where n is the number of FLC output variable fuzzy subsets, c_i is the centroids of the ith fuzzy subset of the FLC output variable, m_{R_i} is the rule activation strength for the ith fuzzy subset of the FLC output variable.

Structure of FLCG

The FLCG package consists of two modules: a program generator (FLCG), and a fuzzy logic function library (FLO). Figure 6-8 shows the structure of the code.

Chap. 6 Fuzzy Logic Software and Hardware

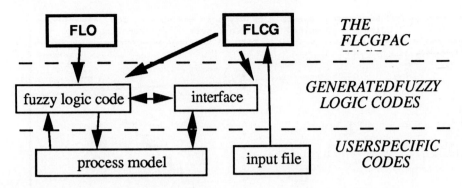

Figure 6-8. The architecture for FLCG software system.

The program generator, FLCG, was compiled to form an application module. It accepts an input file and generates two text files: one is a fuzzy logic program, *control.c*, and the other is an interface module, *glue.h* . The interface module links the fuzzy logic program with the process or plant model. The generated program and the interface module are plant-specific. The input file defines the variable names, their fuzzy subsets, and the fuzzy rules used in controlling the plant. The program *control.c* and the plant simulation model can then be compiled and executed to form a complete fuzzy logic application program.

The fuzzy logic library is represented by the block FLO in Figure 6-8. It implements the fuzzification, rule evaluation and inferencing, and defuzzification functions. This library is used by the generated fuzzy logic code. The defuzzification scheme used is a simplified "centroid method." In this method, the crisp control value is calculated by taking the weighted average of the membership function with the output variable fuzzy subset centroids as the multipliers. The centroid is that point at which the area under the membership function curve to its left equals that to its right. The five-point Gauss-Legendre integration method is used to calculate the area under the curve.

6.1.4 NeuraLogix Fuzzy Microcontroller NLX-230

Another software environment, from NeuraLogix Corporation [4], is called the NLX-230 Fuzzy Microcontroller (NLX-230 FMC) and comes with a hardware card as well as software written in C language. The software package will allow you to set up your own FAM matrix as well as the rule set. The membership functions can also be designed through a staircase-style triangular form, as shown in Figure 6-9.

In NLX-230 FMC the inputs are ranked by how well they fit within a set of user-determined membership functions. The controller uses a linear symmetrical membership function and a simple min/max fuzzy inference scheme for easier digital implementation. Rules are used to determine what set of conditions is present at the inputs. Each rule consists of up to sixteen terms, one for each crisp-input/fuzzy

126 Chap. 6 Fuzzy Logic Software and Hardware

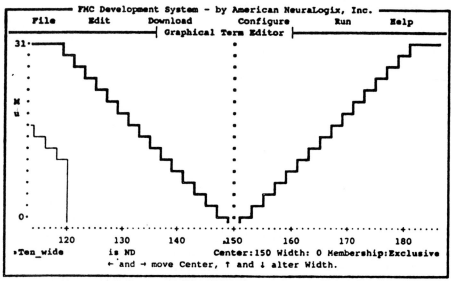

Figure 6-9. NeuraLogix's NLX-230 typical staircase membership functions.

membership function pairing, and by a user-defined Action Value whose amount modifies the output of that rule. The rule which best fits the given set of input conditions determines what modification will be made to the output. Note that both the input ranking and rule processing are performed in parallel for all inputs and outputs. The FMC allows storage of up to 64 rules in the on-chip 24-bit wide rule memory. These rules are shared across all of the outputs. The number of rules available for any one output depends on the remaining number of rules not being used by other outputs.

     The membership function in a fuzzy logic environment often defines the "degree of similarity" between an unknown input and a known value. It has also been shown that the actual shape of the membership function is not as important as the degree of overlap which may exist between membership functions. Moreover, determining the optimum shape of a membership function may not always be easy and can sometimes be obtained empirically at best. The sensor characteristics, control responses, and other dynamic factors often help with the optimum shape of a membership function. To avoid these "black art" approaches to membership function, NeuraLogix Corporation [4] has proposed another scheme for estimating the fuzzy membership function for an incoming crisp input. Instead of noting the intersection point of the input to the membership function and the fuzzy input singleton, the distance from the center of the membership function to the input is measured. This measurement is performed by subtracting the input from the known Center Location and ignoring the sign. Figure 6-10 shows the shape of the membership function used for this calculation. The resulting difference, ac, is then subtracted from the maximum value to obtain d (or m, as in Figure 6-10). If the input and center are the same, ac is zero, and hence d=MAX; the further away from the Center Location, the smaller is d.

Chap. 6 Fuzzy Logic Software and Hardware 127

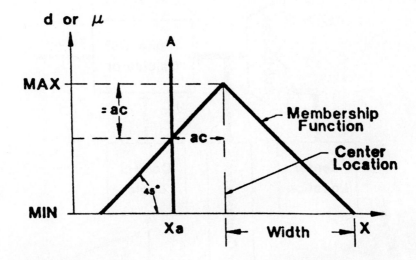

Figure 6-10. Membership function shape used in NLX-230.

Figure 6-11 shows a block diagram of the circuitry used to determine similarity using NLX-230 fuzzy microcontroller [4]. The distance value is compared against the width values in the Comparator. The Control Logic block is used to control data in the Alpha Cut Calculator (see Chapter 2) in the following way: if the distance is greater than the membership function, the result is forced to zero, or min. An exception to this occurs when the Type bit is set so that it causes an exclusive membership function, so if the input is inside, the result would be forced to min. Figure 6-12 shows Inclusive and Exclusive membership functions with Width set at 13. If data is within the limits set by Center Location and Width, the Alpha Cut Calculator outputs the alpha cut result. NeuraLogix claims that this method has two advantages. First, it frees the designer from the difficult decision of membership function determination. Second, the method produces easily implementable solutions to the similarity determination problem.

This method would free the designer from choosing the shape of the membership function, which is somewhat of a "black art" to begin with. It does, however, retain the degree of overlap among membership functions which is an important aspect of fuzzy logic inferencing. It also determines a simple solution of the similarity determination problem [4].

6.1.5 Aptronix's Fuzzy Inference Development Environment - FIDE

One of the more recent fuzzy logic software simulation environments is called FIDE which is a product of Aptronix, Inc. It is a Microsoft Windows-based environment which would allow the user to perform the following tasks : i) Basic functions of fuzzy composition, inferencing, membership function composition, rule set creation and

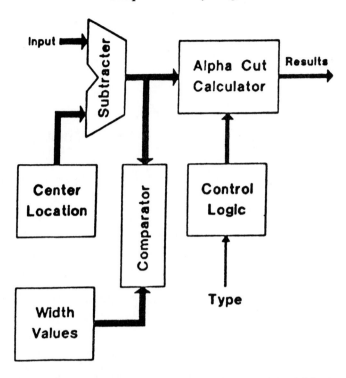

Figure 6-11. Similarity circuitry used in NLX-230.

evaluation., ii) Fuzzy control simulation , iii) Debugging of existing programs, and iv) Code generation for real-time applications.

The basic language in use for FIDE is called FIL (Fuzzy Inference Language) which is a language which describes units in a fuzzy inference system. These units are currently FIU - Fuzzy Inference Unit, FOU - FIDE Operation Unit, and FEU - FIDE Execution Unit. FIL, as with any language such as Togai's Fuzzy_C, has its own syntax and semantics. For example, definition of FIU is based on the mathematical definition of a fuzzy inference unit [5].

Like in FULDEK, FIDE would allow for the user to use self-generating set of membership functions for a given universe of discourse. This is done by a so-called command listSpec as is seen in Figure 6-13. It will also allow the user to create his or her own specific shaped membership function through the MF-Editor/Set Margin operation.

In FIDE, the system variables would appear as block within various windows with appropriate corresponding unit such as FIU, FOU, etc. Using these block, the user can use the Make for Simulation capability of FIDE to create a Composer/Window list which would be used eventually to perform fuzzy logic (control) simulation as shown in Figure 6-14. The defuzzification method used in FIDE are centroid, maximization, left-most maximization, and the right-most maximization defuzzifiers.

Chap. 6   Fuzzy Logic Software and Hardware   **129**

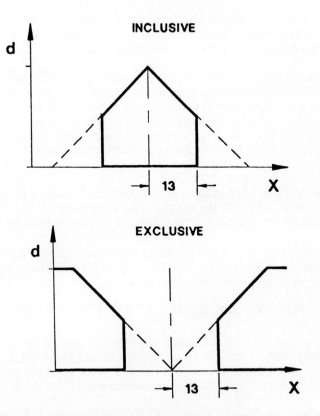

Figure 6-12. Inclusive and exclusive membership functions
used in NLX-230.

In FIDE, the system variables would appear as block within various windows with appropriate corresponding unit such as FIU, FOU, etc. Using these block, the user can use the Make for Simulation capability of FIDE to create a Composer/Window list which would be used eventually to perform fuzzy logic (control) simulation as shown in Figure 6-14. The defuzzification method used in FIDE are centroid, maximization, left-most maximization, and the right-most maximization defuzzifiers.

FIDE has a very appealing 3-D graphical capability. It would allow view of the control surface and tracing capability of a given rule on the surface. It will, like FULDEK, access to the FAM banks. FIDE also has a run-time library which would allow the users a variety of tasks associated with fuzzy logic, fuzzy inferencing, fuzzy decision making, fuzzy control, etc.

FIDE can be used through the following three steps :

1. Source Program - The first step is to write a source program in FIL whereby the input/output variables, their membership functions and fuzzy if-then rules are defined.

In this regards, FIDE is similar to Togai's Fuzzy_C system. Graphics interface is used extensively, however, to assist the user to create its FIL source code. The FIDE compiler then converts the code into the Aptronix standard data structure.

  2. Code Debugging - This stage would allow the user to view the code from multiple perspectives. Three tools are available to the user. These are : *Tracer* in which the user can attach a given input variable to any test value and observe the output. This can be done in step by step fashion similar to FULDEK. *Analyzer* which provides a display of a global view of the transfer function response. *Simulator* which is the final stage of the FIDE's debugger, simulates the unit's dynamic behavior and displays the corresponding plots.

  3. Real-Time Code Generation - Once a processor has been selected, the code can be used for a specific chip. At the present time FIDE supports the Motorola chips MC6805, MC68HC05, and MC68HC11 and plans are for more chips to be included.

  At this writing, FIDE is just becoming available; hence the author has not had the full opportunity to evaluate FIDE completely. Interested readers may obtain the manual of this new fuzzy logic software environment [5].

6.2   SOFTWARE SIMULATED EXAMPLES

In this section we present two simulation experiences. One is the much-discussed "inverted pendulum" using Togai's Fuzzy-C system with MATLAB [6], and the other is the simulation of a thermal system using MATLAB 's .m file capability to create an "m" file in place of NeuraLogix's NLX-230 fuzzy microcontroller.

6.2.1 An Adaptive Fuzzy Control of the Inverted Pendulum

Here, a comparative study of classical expert system and an adaptive fuzzy control system for an inverted pendulum are given. This comparative study was done by means of computer simulation through the MATLAB and Togai InfraLogic's Fuzzy-C software programs [1]. The inverted pendulum, or broom-balance problem, consists of a cart mass M movable in a horizontal direction with a velocity v, and an attached broom with mass m and inertia J, whose angular displacement q and angular velocity W could be detected and measured. The means of balancing the broom on the cart is an external force u that can be applied to the cart in the horizontal direction.

  The mathematical model of an inverted pendulum is well documented [7]; note that it consists of a fourth order nonlinear unstable system. In its linearized form, it has two poles at the origin, one in the right-hand plane and one in the left-hand plane. The system was linearized about the region $-5^\circ <= q <= 5^\circ$ and then a state feedback controller was designed for it. This pole placement was done by a full state feedback such that u=Kx, and K's were determined such that poles were placed at $-2, -1$, and $-1\pm j$. By virtue of the definitions of the system's states, this state feedback controller is effectively a PD controller. The plant was made stable with this controller.

Chap. 6   Fuzzy Logic Software and Hardware   **131**

Adaptive Fuzzy Expert Control System

In an attempt to improve the response of the system, a variable feedback (self-tuning) controller was designed. In order to acquire the knowledge base the state feedback gains were changed during the simulation. It was experimentally discovered that in order to obtain the greatest impact on the stability and response of the system it was necessary to vary only K2 and K3 state feedback coefficients, which actually corresponds to the change of the cart velocity and the angle of the broom. In order to see "real" improvement of the system, the angle q was limited to change $\pm 5^{\circ}$, and the change of the displacement of the cart to ± 0.5 meters, so that the problem could be potentially implemented in hardware on a 1-meter track. The idea behind the expert system design is that if a large displacement was detected in the output of the plant, a big state feedback value would be applied to control the plant, and vice versa. If small q displacement is detected, small state feedback would be applied. With this method, the overshoot of the system would be improved, while the time response of the system would be increased.

Both the constant controller and the expert system controller are "crisp" systems, i.e., both have certain crisp values for the state feedback controller in response to certain angular displacement. With the fuzzy system it is possible to have an adaptive controller for the dynamic system, but this controller will change for all the intermediate values of angular displacement that were not implemented before; moreover, this change will be nonlinear in response to the nonlinear dynamic system.

First, the input and output fuzzy variables were determined. Previously, it was determined that the best control for the system is achieved by adjusting x_2 (cart velocity), and x_3 (angular displacement) states. Therefore, these two states will be our sensor outputs from the fuzzy controller. These outputs are designated as K2FAC and K3FAC, respectively. From previous experimenting with the expert system, limitations on these variables were determined as: q for $\pm 5^{\circ}$; K2FAC and K3FAC 1-2 (this is just the multiplication factor). The purpose of the fuzzy controller is to improve the settling time and the stability of the system. Another advantage of this system is that a previously acquired data knowledge base can be used, except now the mapping for the state feedback elements will be nonlinear.

Next, the membership functions for each variable must be assumed using typical trapezoidals at the ends and a triangular in between the intervals. The membership function for q is evenly spread between +5° and –5° degrees over seven regions: negative big, medium, and small; zero; and positive big, medium, and small. K2FAC and K3FAC membership functions are equal, but unevenly spread over four regions: positive small, medium, big, and wide. The reason for the uneven spread is that we need a lot of compensation when q is ± 3 degrees or bigger, but as q gets smaller, less compensation and more "finesse" are needed in the compensation because we want the system to settle quickly and with the minimum overshoot. Therefore the membership function is denser and more overlapped for small K's so we can achieve a "linear" effect.

Next, determine fuzzy rules for the system. This is done by mapping q membership function into K2FAC and K3FAC membership functions. Therefore, mapping is done as follows:

1. when THETA is PB or NB, K's are mapped as PW (very big compensation)
2. when THETA is PM or NM, K's are mapped as PB (big compensation)
3. when THETA is PS or NS, K's are mapped as PM (medium comp.)
4. when THETA is ZO (close to zero), K's are mapped as PS (small compensation).

Simulation and Results

The simulation of the expert and fuzzy expert control systems were done by interfacing MATLAB and Togai InfraLogic's Fuzzy-C expert system environment. Figure 6-13 shows the simulation approach. The Togai's fuzzy controller was used in conjunction with MATLAB's differential equation solvers (ode45.m) capabilities to complete the fuzzy control systems simulation. The former environment is used to simulate the dynamic behavior of the system, while the latter represents an environment for the fuzzy controller (rules, membership function conflict resolution, etc.). The connection between the two simulation environments was made through the C-language shell which allowed MATLAB and Fuzzy-C to communicate. The system's simulation was done in three stages: 1) Constant feedback control system, 2) Expert control system, and 3) Adaptive fuzzy expert control system.

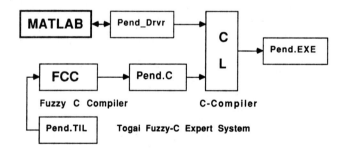

Figure 6-13. A combined Fuzzy-C expert system and MATLAB simulation.

A typical simulation result is shown in Figure 6-14. The time response for angular position is shown here for all three controllers. It is noted that under the expert system controller, both overshoots and settling times consistently worsened for the angular displacement. The third simulation was done when fuzzy IF-THEN rules and the membership functions were used to tune the controller's gains. Both the overshoots and settling times improved. Also shown in Figure 6-14 are the usual smoothing and interpolation effects of fuzzy rules in between crisp expert rules. It is further noted that both the overshoot and settling time have been improved through the fuzzy expert controller. The comparison between these two controllers is also in favor of the fuzzy

Chap. 6   Fuzzy Logic Software and Hardware   133

controller. Both settling time and overshoot have been improved, but the rise time has increased by about 4 seconds for the fuzzy controller. Overall, the results of the simulation compare the fuzzy controller very favorably with both constant-gain (pole placement) and expert (crisp logic) controllers.

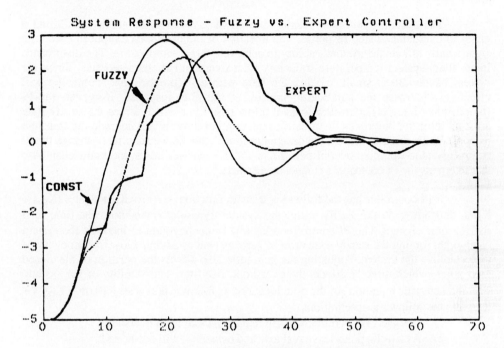

Figure 6-14. Simulation results of three control schemes for the inverted pendulum.

6.2.2 A MATLAB Approach to NeuraLogix Fuzzy Control of a Thermal System

In this section MATLAB [6] is used to create a macro file (.m file) to simulate NeuraLogix Corporation's fuzzy microcontroller [4] for a thermal system which is an approximation of an air conditioning system. In the thermal system, the transfer function relates the process temperature y(t) to the input control m(t) through the thermal dynamic properties of the chamber. The variable d(t) models the random disturbance of the door opening to the chamber.

The transfer function of the above system is assumed as follows: $G(s) = Y(s)/M(s) = 1/(s+0.05)$, with $M(s) = 1/s$; and $D(s) = -0.5/s$. A proportional plus integral controller is commonly used to track the temperature of the chamber such as the inside channel of an air-conditioner. In this section a fuzzy PI controller is designed to simulate the behavior of the air conditioning system.

First, to acquire a knowledge base regarding the plant characteristics, the system was run through a series of simulations using a variety of disturbances: 1) no disturbance, 2) a continuous disturbance of 1, 3) a delayed disturbance, and 4) a pulsed disturbance,. Due to lack of space most of the simulation results are not presented here. More details can be obtained from Reference [8].

The system simulation results showed the system to be stable, exhibiting a classical overdamped step response. The injection of the disturbance input in each case significantly affects the system's ability to achieve a target temperature. The disturbance input alters system temperature trajectory with overdamped characteristics. Since the system bandwidth is small, closing the loop with a P (proportional) controller will effectively increase the bandwidth, improving transient response. However, the PI (proportional-integral) controller is itself a low pass filter which forces the steady state error to zero; this improves disturbance rejection but may in fact degrade the transient response in the process. Hence, the design question "How do we adaptively change the PI control gains to optimize overall performance?" The answer lies in the knowledge base that interprets the PI controller's frequency response.

A PI controller has the following transfer function: $Gc(s) = K_p + K_i/s = (K_p \cdot s + K_i)/s$, depending on the K_i/K_p ratio, the system dynamics-overshoot, rise time, and settling time change. The PI controller adds low frequency gain to increase the system bandwidth, forcing the steady state error to zero, as well as adding a low frequency phase to destabilize the system. Adjusting the gain ratio also affects the position of the closed loop poles, which directly affects the overshoot, rise time, and stability of the system. The characteristic equation set the pole locations as follows: $p(s) = s^2 + (0.05 + K_p) \cdot s + K_i$. Note the following observations:
1) If $K_p > K_i$ poles remain on the real axis, there is no overshoot;
2) As $K_p < K_i$ poles leave real axis and overshoot will occur; and
3) As K_p is set larger than K_i the dominant pole approaches the origin leading to a slower transient response.

With this frequency and root locus information to form a preliminary knowledge, the next task is to address the simulation model governing the system, with attention to the details of control resolution, and finite computation precision. Figure 6-15 shows a block diagram of an adaptive fuzzy PI control for the thermal system. The system has the following numerical description: Sensor was assumed to be represented by 0.05 v/Co, DAC by 4096 bits per 5 volts, ADC by 24 volts per 4096 bits, and set point by 40 degrees C.

In facing the issue of control, fuzzy or non-fuzzy ,the degree to which the outputs can be measured sets the limitations on how well the system will eventually be controlled. Since most control and all adaptive/algorithm-based control is done digitally there is a need to quantize the effects of digitization.

In choosing the rule set/membership functions for the adaptive PI controller, the previous information regarding the pole locations and frequency response was considered. Figure 6-16 shows the membership functions of the absolute error for time period n and n-1 to be used for the subsequent simulation.

Chap. 6 Fuzzy Logic Software and Hardware

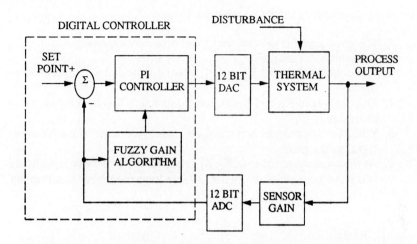

Figure 6-15. Schematic of an adaptive fuzzy PI control of a thermal system.

The absolute value of the error signal is used because the input must be between 0 and 255. No membership functions are given for the outputs K_p, and K_i since the NeuraLogix chip does not fuzzify the outputs. Below are the fuzzy rules used here :

RULES:
 OUTPUT 1
  If ABS(error(n)) is SMALL and ABS(error(n-1)) is SMALL
  then $K_p = 12$

  If ABS(error(n)) is SMALL and ABS(error(n-1)) is MED
  then $K_p = 18$

  If ABS(error(n)) is SMALL and ABS(error(n-1)) is BIG
  then $K_p = 24$

  If ABS(error(n)) is MED and ABS(error(n-1)) is BIG
  then $K_p = 16$

  If ABS(error) is MED then $K_p = 10$

  If ABS(error) is BIG then $K_p = 8$

 OUTPUT 2
  If ABS(error) is ZERO then $K_i = 10$

  If ABS(error) is SMALL then $K_i = 12$

  If ABS(error) is MED then $K_i = 8$

The following observations should now be noted:

1. The error signal is the only input that determines outputs.
2. All outputs are integers.
3. The output values for K_p and K_i change in unison since they use the same input.
4. The changes in K_p and K_i are always such that $K_p > K_i$, keeping the poles on the real axis.
5. When the error signal is large K_p/K_i values are large to increase the control signal to the plant.
6. As the error signal reduces, K_p/K_i values reduce to prevent overshoot.
7. When the error signal is small K_p/K_i are large for increased sensitivity

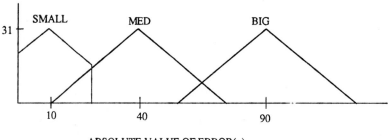

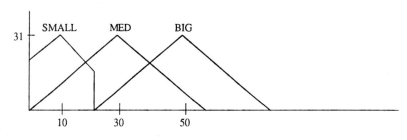

Figure 6-16. Membership functions for the absolute errors for the thermal system.

The following simulation made use of the error signal at time (n) and the error signal at time (n-1), in order to provide more information to the fuzzy rule base and thus improve results. The simulation was performed using the membership function from Figure 6-16 and the fuzzy rules just presented. Figure 6-17 shows the new step responses of the thermal system using both fuzzy and crisp PI controllers.

The simulation shows that the fuzzy controller does reject disturbances more smoothly, but the crisp PI is fairly adequate as well.

Chap. 6   Fuzzy Logic Software and Hardware   **137**

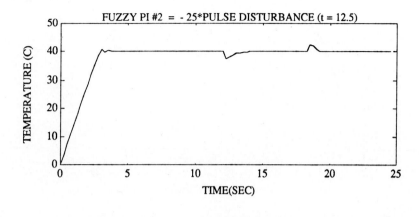

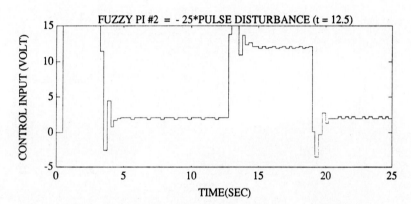

Figure 6-17(a). Fuzzy PI control of the thermal system.
Compare with Figure 6-17(b).

6.3.   FUZZY LOGIC HARDWARE

The recent interest in fuzzy logic can best be attributed to the creation of fuzzy chips in the mid-1980s. Four developers were among the first to produce the chips. These individuals, who were all Japanese and Japanese-Americans, developed the various types of chips summarized in Table 6-7.

The basic notion behind fuzzy chips is the so-called "fuzzy cure fitting" or interpolation, or "interpolative reasoning" (see Figure 6-18). It is noted that each crisp point is replaced with a rectangle of points (patches) and these clusters may overlap. For a given value of X, the value of Y is determined by noting which cluster is the value X nearest. If an X value is equidistant from two or more clusters, its resultant Y value can b be made proportional to the output values suggested by all of the clusters of which X is partially a member. Thus a fuzzy chip provides a built-in approach to the mapping or

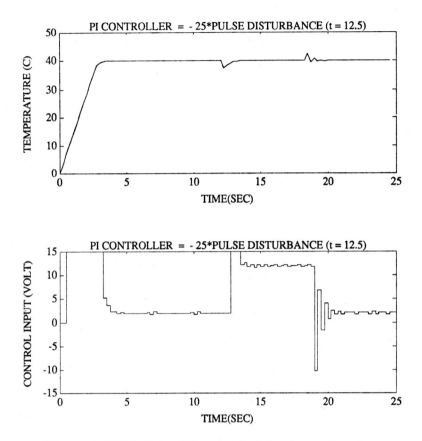

Figure 6-17(b). Crisp PI control of the thermal system. Compare with Figure 6-17(a).

NAME	SPEED (FIPS*)	NO. OF IN/OUT	NO. RULES	TYPE
Togai	400K	256/128	1000	Digital
Wanatabe	625K	4/2	51	Digital
Hirota	6M	2/1	Unlimited	Digital
Yamakawa	200K	3/1	1	Analog

* Fuzzy inference per second

Table 6-7. Early development of fuzzy logic chips.

Chap. 6 Fuzzy Logic Software and Hardware

interpolation problem. It is noted that in more recent time, many other fuzzy chips either have come or soon will be coming into the market. Notable among these are Omron Corporation and NeuraLogix as well as Motorola, Inc. Space and time are not available here to give a detailed discussion of these chips, however two fuzzy cards which have been used at University of New Mexico's projects will be discussed below: the Togai Single-Board Fuzzy Controller and the NeuraLogix Fuzzy Microcontroller.

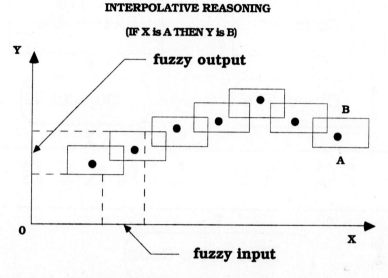

Figure 6-18. The notion behind fuzzy chips: Interpolative Reasoning.

6.3.1 Togai Single-Board Fuzzy Controller

The single-board fuzzy controller (SBFC) comes with the standard Togai fuzzy chip (see Table 6-7) and its own A/D and D/A chips, as well as the 230 standard BUS and other interface. It has a 1-Mbyte erasable and electrically programmable ROM (EPROM). Software is available which translates the fuzzy expert control system developed by Togai Fuzzy-C into machine code for the microprocessor used on the board. Over 800 rules can be supported in the hardware and up to 200,000 fuzzy rules can be evaluated in one second. The programming procedure for the SBFC is similar to that of programming in Fuzzy-C on a PC. There are, however, some very distinct differences. Programming in Fuzzy-C on a PC, the standard Fuzzy-C code can be used as a debugging tool and a check on the hardware implementation. In using SBFC the fuzzy programming language (FPL) code is compiled directly to an object code. The driver code (main program) is written in FC110 assembly code instead of C, as in the Fuzzy-C system. The FC110 assembler is used to convert the driver to an object code, and the FC110 linker can then be used to combine the two object codes into a code that is executable by the FC110 digital fuzzy processor (DFP). This procedure is described in Figure 6-19.

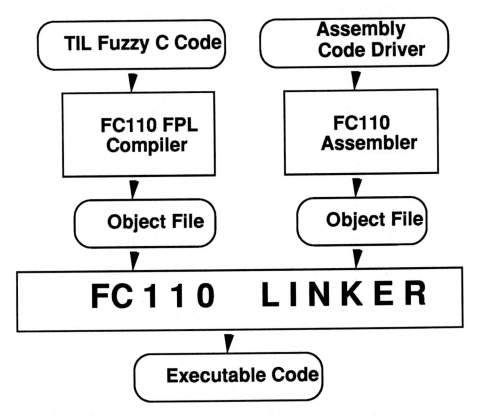

Figure 6-19. Steps for creating hardware ready fuzzy control program.

It is possible to interface the SBFC board with a personal computer. For very complicated control programs this may be the best way to develop and debug the codes. Results of a recent project by Shao and Parkinson [9] showed that it is not necessary to build an interface if the user-developed assembly code is used only for interacting between the hardware components on the board and the fuzzy rule codes. Since fuzzy rule bases can be easily developed with the aid of the Fuzzy-C development systems software mentioned earlier, a complicated real-time fuzzy control system can still be implemented using the code described here. The following steps summarize the programming of Togai's SBFC:

1. Set public and external symbols
2. Initialize the stack pointer used for calling subroutines
3. Load sensor input from A/D converter address to the fuzzy rule routine input variable address.
4. Call the fuzzy rule routine
5. Load control output from the fuzzy rule routine output variable address to the D/A converter address.
6. Go to Step 3.

Chap. 6 Fuzzy Logic Software and Hardware 141

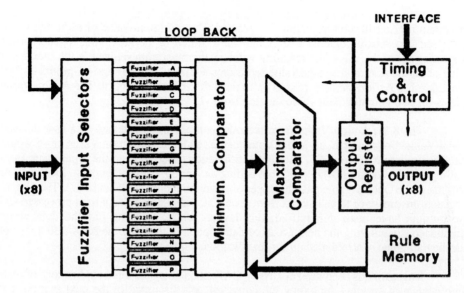

Figure 6-20. Block diagram for NLX-230 fuzzy microcontroller.

In a later section Togai's SBFC is used to control a conveyor belt system in robotic assembly application.

6.3.2 NeuraLogix Fuzzy Microcontroller

In an earlier section the NeuraLogix NLX-230 fuzzy microcontroller was briefly introduced. Here a few more words will be given regarding the card itself. Figure 6-20, a block diagram of the NLX-230 FMC, shows that there are 16 fuzzifiers associated with this controller. Each of these fuzzifiers has a 1-of-8 selector which allows each fuzzifier to input data from any one of the eight inputs. Each fuzzifier looks at only one input. If a fuzzifier is configured to be used in a Loop Back from an output, then it cannot be used to process an external input. This conjunction of an input with a membership function performed by a Fuzzifier is called a "Term" by NeuraLogix Corporation [4]. The NLX-230 supports a maximum of 16 such "Terms."

Each fuzzifier, using the similarity notion discussed earlier, calculates the value of the membership function for its input signal. There is an 8-bit Center Location associated with each Fuzzifier. Membership values which fall within the Width (see Figure 6-10) are passed onto the Minimum Comparator for rule processing.

The Minimum Comparator block is implemented with a Neural Network for high-speed throughput. The first 16 bits of the 24-bit rule are used to enable the outputs of each fuzzifier to compare the minimum membership values. Each rule bit has a fixed relationship with a fuzzifier. When a rule bit is asserted "1," the corresponding Fuzzifier outputs are compared with one another to find the minimum value.

Once the minimum value for a rule is found, it is stored in a temporary maximum register in the Maximum Comparator. The minimum value of each rule being processed is compared against a reference value stored in the maximum register. After processing all of the rules, the resulting value in the maximum register is the maximum value of all the minimums. This value represents a "fuzzy" logical sum-of-products. The winning rule's Action Value is passed on to the defuzzification control. The Action Value is an 8-bit two's complement value ranging from -128 to 127.

The 8-bit Action Value from the rule memory word is added to a user-defined Initial Value. For example, an Initial Value of 100 added to an Action Value of –5 would result in an output value of 95. Saturable arithmetic is used to keep the input value from "rolling over" (modula 128) in the event adding an Action Value attempts to increment or decrement the Output Value past the upper or lower boundary (-128 and 127). The Output Registers are pipelined to enhance performance. Once all the rules have been processed and the outputs have been defuzzified, the device begins clocking out the data while simultaneously entering the next group of data inputs. In this way the NLX-230 does not use the traditional centroid method for fuzzification.

This card has been used at UNM's CAD Laboratory for three applications of real-time control of systems, including the conveyor belt discussed in the next section, but results were not available in time for inclusion here. Future publications will incorporate NLX-230's real-time implementations.

6.4 HARDWARE EXPERIMENT EXAMPLES

This section will briefly discuss two real-time fuzzy controller implementations designed in a laboratory setting. Currently, four real-time control architectures are being implemented for real-time fuzzy control systems. Figure 6-21 gives a pictorial presentation of all four implementations. As shown, three are based on Togai InfraLogic [1] software and hardware and the fourth is based on NeuraLogix's FMC board [4].

6.4.1 A Conveyor-Belt Experiment - Fuzzy-C Approach

Figure 6-21(a) shows the experimental set-up for the control and tracking of a robotic station's conveyor belt using self-tuning fuzzy PI controller. The fuzzy controller was implemented on a Packard Bell 386 personal computer, which acted as host for the Togai's Fuzzy- C expert system. A user-interactive program was written to prompt the user with data for the fuzzy controller implementation. Fuzzy IF-THEN rules were used to automatically adapt the gains of the PI controller in reaction to system disturbances and unmodelled dynamics. For example, one experiment dealt with various load conditions and sudden load changes on the conveyor belt. The conventional controller acted relatively satisfactorily for low load, but the fuzzy controller could adjust promptly for any load situation, including high load. Typical rules were :

```
          IF ERROR IS NB OR PB THEN GAIN IS PB
          ELSE
          IF ERROR IS NS OR PS THEN GAIN IS PS
          ELSE
          IF ERROR IS ZERO THEN GAIN IS PM
```

Chap. 6   Fuzzy Logic Software and Hardware   **143**

The corresponding membership functions for GAIN and ERROR were chosen as standard trapezoidals at the two ends of the interval and an overlapping triangular in between intervals. Figure 6-22 shows the error behavior for load and no-load conditions using PI (crisp) and fuzzy PI controllers. Although this system was fairly linear and well-behaved, and the true nature and power of fuzzy logic cannot be tested, it did allow the authors to realize a real-time control implementation.

6.4.2 A Conveyor-Belt Experiment - SBFC Approach

In this section the conveyor belt system of the previous section is controlled using the Togai's SBFC card, which was also described earlier. The goal of the experiment remained that of controlling the speed of a DC motor driving a conveyor belt. The speed of the belt will vary under load and no-load conditions. The sensed input is the speed of the motor in the form of a voltage from a tacho-generator. The rules base for the hardware implementation are given by the following set of four rules [9]:

1. IF the input voltage is VLOW (very low) THEN output voltage is HIGH
2. IF the input voltage is LOW THEN output voltage is MEDIUM
3. IF the input voltage is MEDIUM THEN output voltage is LOW
4. IF the input voltage is HIGH THEN output voltage is VLOW

The input and output voltage membership functions are shown in Figure 6-23. For verification purposes the input/output rules for the fuzzy controller were simulated on the Fuzzy-C system. The resulting input/output of the fuzzy controller using simulation is shown in Figure 6-24. Then the SBFC implementation of the conveyor belt, as shown in Figure 6-25, was performed using the steps outlined in the section headed "Togai Single-Board Fuzzy Controller." The resulting hardware implementation of the input/output of the fuzzy controller is shown in Figure 6-26. Note the close correspondence between this figure and the curve in Figure 6-24. Disregard the discrepancy in the scaling of the axes of the latter figure; this was done for ease of plotting, and the two scales are in fact the same.

6.5   CONCLUSIONS

Since 1989 many software and hardware projects have been undertaken within the CAD Laboratory at the University of New Mexico, where all of this book's editors and authors have worked. The projects discussed here represent significant progress in the applications of fuzzy logic to control, pattern recognition, digital filtering, and other areas. The details of these projects appear in Chapters 7 through 18.

One important project, which at this moment is not quite ready for reporting, is a fuzzy PI Control of a non-CFC based air-conditioning system. This project, developed in collaboration with local industry, attempts to implement an adaptive fuzzy PI controller for a newly patented non-freon based air conditioning system for trucks and buses. The controller will be based on the NeuraLogix fuzzy microcontroller (Figure 6-21[d]).

We conclude by once again noting that it is difficult for a single chapter to cover the wide spectrum of software and hardware currently available in the marketplace. We hope the reader will take this chapter as an initial reference for further reading and research into this important aspect of fuzzy logic and fuzzy control.

ACKNOWLEDGMENTS

The writing of this chapter would not have been possible without the remarkable work of so many of the author's capable students. Thanks to Mr. Denis Barak, Mr. Narul Rashid-Alang, Mr. Steve Baugh, Mr. Jinqiu Shao, and Dr. Jerry Parkinson for their many dedicated hours of work. The collaboration of Mr. Mark Dreier of Bell Helicopter Textron is also sincerely appreciated. Finally, deepest gratitude goes to Dr. Lotfi Zadeh, the father of fuzzy logic, for his generous encouragement.

REFERENCES

[1] Togai InfraLogic, Inc. *Fuzzy-C Expert Systems User's Guide*, 1990, Irvine, CA, 1990.

[2] M. Dreier. *FULDEK - Fuzzy Logic Development Kit User's Manual*, Bell Helicopter Textron, Fort Worth, TX, 1992.

[3] N. Rashid-Alang. *FLCG - A Fuzzy Logic Code Generator*, Technical Report. Department of Chemical and Nuclear Engineering, University of New Mexico, Albuquerque, NM, 1992.

[4] NeuraLogix Corporation. *User's Guide for NLX-230 Fuzzy Microcontroller*. Sanford, FL, 1991.

[5] Aptronix Corporation. *User's Guide for F.I.D.E Fuzzy Inference Development Environment*, Palo Alto, CA, 1992.

[6] Mathworks, Inc. *MATLAB User's Guide*. Boston, MA, 1991.

[7] M. Jamshidi, M. Tarokh, and B. Shafai. *Computer-Aided Analysis and Design of Linear Control Systems*. Prentice Hall Publishing Company, Englewood Cliffs, NJ, 1992.

[8] M. Jamshidi, R. Marchbanks, K. Bisset, R. Kelsey, S. Baugh, and D. Barak. "Computer-Aided Design of Fuzzy Control Systems," in *Advances in Computer-Aided Control Systems Engineering*, M. Jamshidi and C. J. Herget (Eds.). North- Holland Publishing Company, Amsterdam, 1993.

[9] J. Shao and G. Parkinson. "Programming and Testing of the Togai Single-Board Fuzzy Controller," Internal Report. CAD Laboratory for Intelligent and Robotic Systems, University of New Mexico, Albuquerque, NM, 1992.

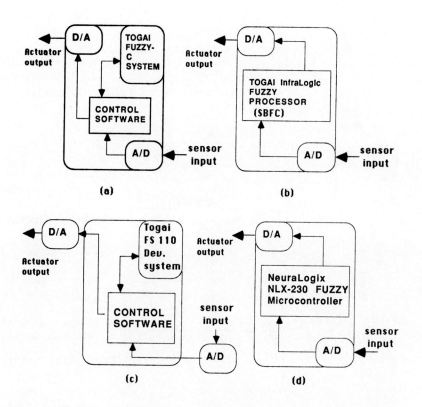

Figure 6-21. Four real-time architectures for fuzzy control systems: (a) Software Implementation (Fuzzy-C); (b) Togai SBFC; (c) Togai FS110 System; and (d) NeuraLogix's NLX-230 FMC.

145

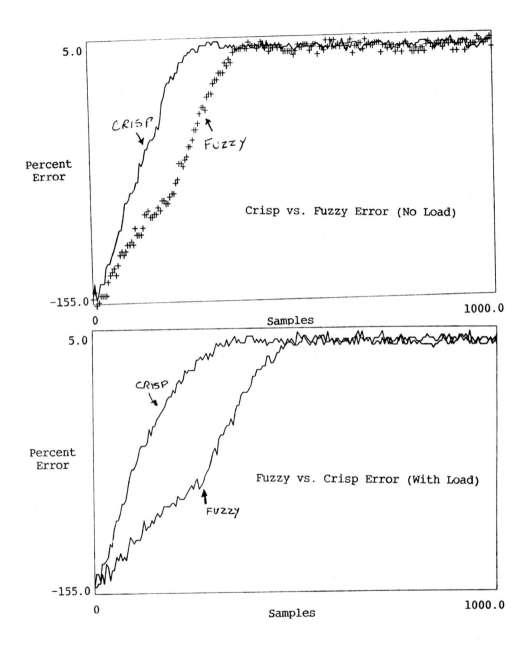

Figure 6-22. Experimental results for fuzzy PI control of a conveyor belt.

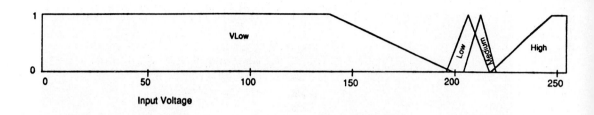
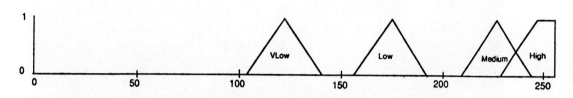

Figure 6-23. Membership functions for input/output voltages in SBFC experiment.

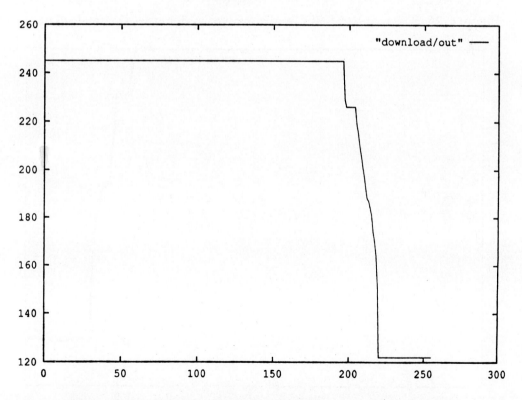

Figure 6-24. Software simulation for the input/output voltages for conveyor system.

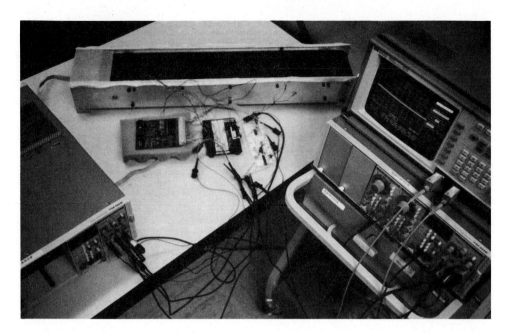

Figure 6-25. Schematic of SBFC and conveyor system experimental set up.

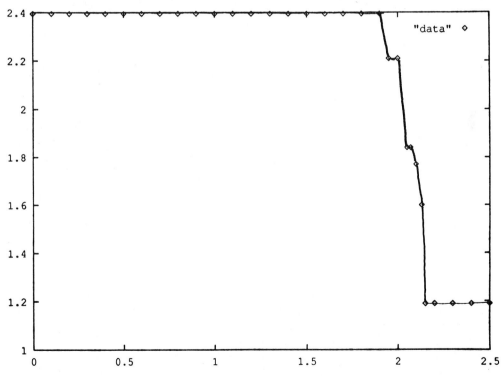

Figure 6-26. Hardware result for conveyor system input/output voltages.

7

A FUZZY TWO-AXIS MIRROR CONTROLLER FOR LASER BEAM ALIGNMENT

Richard D. Marchbanks
University of New Mexico
(Current address: Los Alamos National Laboratory, Los Alamos, NM 87545)

Fuzzy logic was used to control a two-axis mirror gimbal for aligning a laser beam using a quadrant detector. A PC based controller was configured with a simple rule set to control and stabilize the device. A comparison between a non-fuzzy proportional controller and the fuzzy controller was made. Experiments were performed to study the tracking characteristics and the effect of rule pruning on the fuzzy controller response.

7.1 INTRODUCTION

The use of fuzzy logic is rapidly becoming a popular method of solving problems that require a human-type response. Fuzzy logic uses sets of linguistic variables to solve problems where information is imprecise or a mathematical model of the system is difficult to obtain. Fuzzy logic has many applications such as pattern recognition, clustering, data fitting, diagnostics, and the most prevalent being control.

In the case of control, many systems are very nonlinear in nature and obtaining models of these systems are not only difficult but can be impossible. Many systems are described by a linear model that approximates its behavior over a small region. For such systems, the model only works well in this small region. Outside this region, the systems behavior can become unpredictable or unstable. By using of fuzzy logic, one can easily realize such nonlinear systems without the requirement of a model.

7.2 PROBLEM FORMULATION

Communication systems are desired that use line-of-sight transmitting and receiving. This is especially true for submarine communication. A submarine requires the ability to communicate with other platforms but still remain "invisible." A major requirement for such a system is the ability to maintain a communication link between platforms given a variety of disturbances such as movement, bumps, atmospheric or hydrospheric perturbations, and obstructions.

This chapter presents a demonstration of fuzzy control applied to such a problem. It was an attempt to acquire educational experience in a fuzzy logic control and not breakthrough research in solving the proposed problem. Presented are the results of work in a one semester graduate course in fuzzy logic control offered by the department of Electrical Engineering and Computer Engineering at the University of New Mexico.

7.3 APPROACH

A two-axis mirror gimbal is set up to steer a laser beam onto a quadrant detector where the beam spot is desired to lie on the center of the detector. Electronics sense the error in the position of the beam relative to the center of the detector and produces two signals representing the x and y direction errors. These errors are converted to a digital format for use by the personal computer driven fuzzy controller. The controller processes the error information using fuzzy logic and chooses velocity signals for the motor driver. The driver supplies the required power to run the motors that repositions the beam. Figure 7-1 outlines this closed-loop system.

Chap. 7   Fuzzy Laser Beam Alignment   151

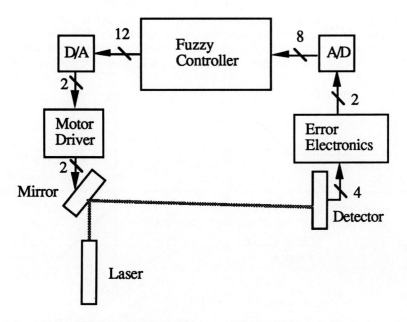

Figure 7-1. Block diagram of closed-loop system.

7.4   HARDWARE

For feedback information, a quadrant detector was set up to sense the error in the position of the beam from the desired position. The center of the detector is where the beam center is desired. A quadrant detector is simply four photosensitive semiconductor elements that produce voltages proportional to the amount of photon flux incident on each one of them. A centered beam spot will have one quarter of its light on each quadrant and the potentials will be equal. As the beam moves off-center, the proportions of the light on each quadrant change, causing their potentials to change. The differences in the potentials are used to determine the direction and magnitude of error in the beam spot's position. Error in the horizontal and vertical directions can be separated by some electronics to reduce the problem into 2 one-dimensional ones. Given the size and price of commercially available quadrant detectors, a decision was made to build one using solar cells. This resulted in a detector with a 16 square cm area and cost under 10 dollars.

Beam steering was achieved with a mirror mounted on a movable gimbal with horizontal and vertical degrees of freedom. Positioning was accomplished by 2 dc motors mounted on the mirror. A complication with using dc motors is that they have a nonlinear speed as a function of armature voltage. The main problem lies in the *dead band* region where the motor requires a minimum voltage to begin moving. Any voltage below this turn-on voltage will not drive the motor. This nonlinear characteristic is shown along with the required control response in Figure 7-2. A driver employing this response will always keep the armature voltage above the turn-on voltage when there is non-zero error. When there is a very small error, the armature voltage, Va, will be at or above the turn-on voltage of the motor. This response however, adds instability and can cause the system to limit cycle when the error voltage is very small. Limit cycling is due to the very high slope of the control response curve near the origin; the motor switches between forward and reverse directions. Part of the objective in the design was to reduce or eliminate limit cycling with the fuzzy logic controller without the requirement of stepper motors.

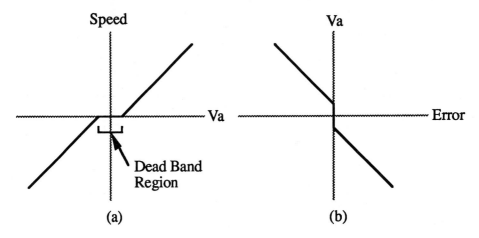

Figure 7-2. a) Motor characteristics; b) Desired armature voltage response.

The controller was employed using a PC with a 386 microprocessor. A Keithley-Metrabyte 8-bit analog-to-digital converter was used to put the analog error in the digital format required by the computer. The input of the A/D converter has a range from −5 volts to 5 volts with a resolution of 8-bits over this range. The error signal was conditioned with amplifiers to fill this range and acquire the full 8-bit resolution. After processing, the resulting digital response signal was converted to an analog voltage for driving the motors by a Keithley-Metrabyte 12-bit digital-to-analog converter.

7.5 SOFTWARE

Designing the required software to perform fuzzy inference can be a very involving task but fortunately there are a few packages that have been written to do this. One of these packages, Togai InfraLogic Fuzzy C, was used to write the primary code that performed the fuzzy arithmetic. This software is programmed in a language developed by the manufacturer that allows the user to easily construct a subroutine containing membership functions and a rule base. The resulting code is compiled into C by a special compiler. The compiled subroutine performs all the required fuzzy operations, including defuzzification, and produces a crisp output. A driver program was written to communicate with the A/D and D/A converters and the fuzzy subroutine. It also was used to acquire data from experiments. Listings of the Fuzzy C code, the A/D and D/A converter addresses, and driver program are listed in appendices A, B, and C respectively.

To represent the error input to the controller, a set of linguistic variables was chosen to represent 5 degrees of error, 3 degrees of change in error, and 5 degrees of armature voltage. Membership functions were constructed to represent the input and output values' degrees of truth for each of these linguistic variables. These membership functions are shown in Figure 7-3.

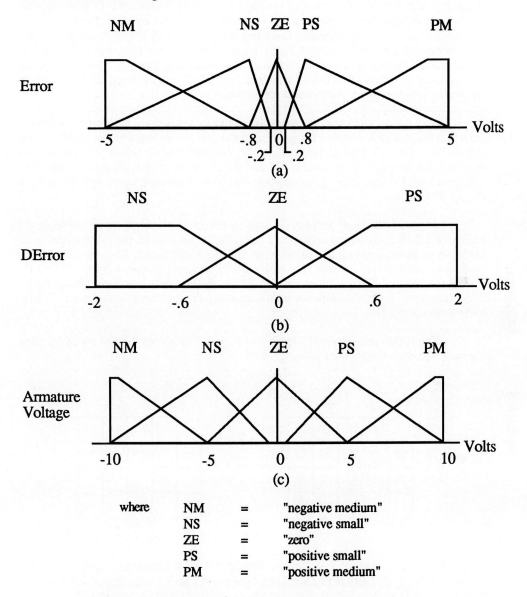

where NM = "negative medium"
       NS = "negative small"
       ZE = "zero"
       PS = "positive small"
       PM = "positive medium"

Figure 7-3. Membership functions. a) Input error,
b) Input change in error, c) Output armature voltage.

The input error membership functions contain a very narrow zero error (ZE) region. A faster convergence was expected by doing this because for a small error, the degree of "medium" will be large and the armature voltage will correspondingly be larger than if ZE were broad. Also, the "medium" membership functions for both the error and armature voltage were chosen not to intersect at zero. This was done to help suppress overshoot and limit cycling. because the motors response was equal in the forward and reverse directions, the membership functions were made with a symmetry about zero. Here is one of the advantages of fuzzy logic. If the motors response functions are asymmetrical, having a different response in the forward direction than the reverse direction, the membership functions can be designed to make the response act as though the motors functions have symmetrical responses.

Two sets of rules were chosen. These "Fuzzy Associative Memories" or FAMs, are a shorthand matrix notation for presenting the rule set. A linguistic armature voltage rule is fired for each pair of linguistic error variables and linguistic change in error variables. For example, if the error is PM and the change in error is ZE then the armature voltage is NM. Because overlap between the fuzzy variables exist, more than one rule can fire simultaneously.

A set of "pruned" rules was used to investigate the effect of reducing the processed information on the controller's behavior. When pruned, the FAM was slightly modified to incorporate all the rules. Note that the FAM banks, like the membership functions, are also symmetrical. The effect on the system's response by modifying the FAM bank is more dramatic than modifying the membership functions. Changing the FAM "coarsely" tunes the response while adjusting the membership functions "finely" tunes the response.

The FAMs are shown in Figure 7-4. Figure 7-4a shows the full set of 15 rules. Figure 7-4b shows the set after pruning.

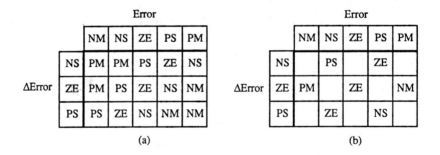

Figure 7-4. Fuzzy Associative Memories.
a) Full set (15 rules), b) Pruned set (7 rules).

7.6 EXPERIMENT

Two experiments were performed to test the integrity of the controller as illustrated in Figure 7-5. First the center finding ability was examined by locating the beam on the detector with the center of the beam offset in the positive x direction from the center of the detector. The system was started with data taken at 100 Hz for 4 seconds. The rate of convergence to the detector center was observed as well as the overshoot and oscillation. This experiment was performed for a non-fuzzy proportional feedback controller, the fuzzy controller with 15 rules, and the pruned fuzzy controller with 7 rules.

The second experiment tested the systems tracking ability. The detector was mounted on a linear positioner with the beam centered on the detector. The positioning stage was driven at 0.2mm/s in a direction transverse to the beam incidence. This speed was due to the limit of the hardware used. The distance from the mirror to the detector was 1.5 m. Again, data was taken at 100 Hz for 4 seconds during the experiment. A non-fuzzy proportional feedback controller, the fuzzy controller, and the pruned fuzzy controller were tested using this scheme.

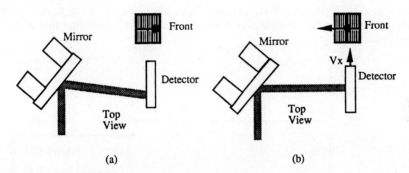

Figure 7-5. Experimental arrangement: a) Center finding, b) Tracking.

7.7 RESULTS

Six experiments were performed to analyze the systems. For center finding ability, a non-fuzzy controller, a fuzzy controller and a pruned fuzzy controller were utilized. These 3 controllers were again used in the tracking experiment. Results are shown in Figures 7-6 through 7-11. Proportional feedback (non-fuzzy) control is denoted by pf, fuzzy feedback is denoted by f, and pruned fuzzy feedback is denoted by pf in the Figures.

Center Finding

With the beam center initially offset in the positive x direction, the proportional feedback controller, the fuzzy controller, and the pruned fuzzy controller were all used to drive the system. The results for 4 seconds of time are shown in Figures 7-6 and 7-7. When comparing the fuzzy controllers to the non-fuzzy controller, the y-error is omitted to prevent the graphs from becoming too crowded.

Graph 6 shows the behavior of the non-fuzzy controller. Both x-error and y-error are shown for 4 seconds of time. The error in the x direction starts at the maximum of 5 volts and remains there until the beam begins to overlap the center of the detector. It quickly converges to zero. The error then overshoots and oscillates as the motor reverses its direction. The y-error remains fairly stable until the beam reaches the center. The oscillation in this error is due to the misalignment of the 4 quadrants of the crudely made detector. When the beam crosses over to the remaining quadrants, the abrupt change due to misalignment causes the oscillation in error.

With the fuzzy controller engaged, the overshoot is dramatically reduced and oscillation is effectively eliminated. This plot is shown in Figure 7-7. Note that the rate of convergence here is nearly the same as the rate of convergence of the proportional (non-fuzzy) controller. This may be due to the speed limitation imposed by the maximum armature voltage.

Next, the pruned fuzzy set of rules was put into use. The plots in Figure 7-8 show these graphs. More overshoot and oscillation are observed in the pruned controller results than in the unpruned controller results but they are better than that of the non-fuzzy controller.

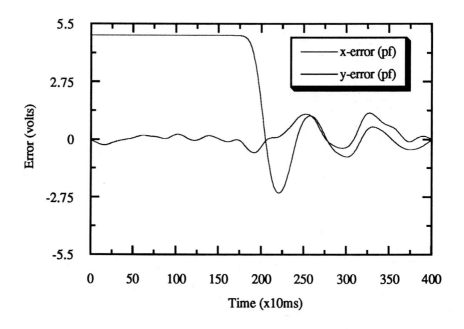

Figure 7-6. Center finding with non-fuzzy proportional feedback.

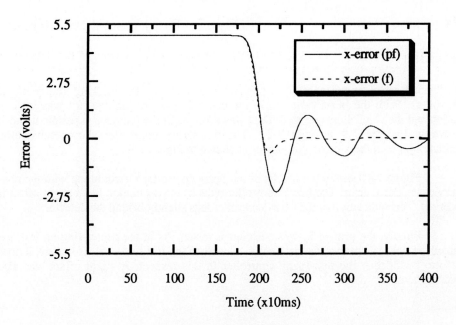

Figure 7-7. Center finding with non-fuzzy proportional feedback and fuzzy feedback.

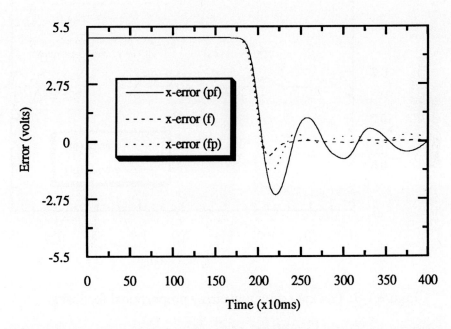

Figure 7-8. Center finding with proportional feedback, fuzzy feedback, and pruned fuzzy feedback.

157

Tracking

The second set of experiments was performed to examine the tracking ability of the controllers. With the beam centered in the desired position, the detector was set into motion and the controllers engaged. Data was taken as in the previous experiment and is shown in Figures 7-9 through 7-11. The non-fuzzy controller shows substantial oscillation in both the x-error and y-error as shown in Figure 7-9.

Figure 7-10 shows the results of the fuzzy controller x-error along with the non-fuzzy controller x-error. The fuzzy controller reduces the oscillation but a slight offset is observed. This indicates that the fuzzy controller lags slightly behind the detector.

Finally, the pruned fuzzy controller is tested. As in the center-finding test, the pruned set does not reduce the oscillation as well as the full fuzzy set. However, it does perform better than the non-fuzzy controller in this respect. A slight offset was also observed here.

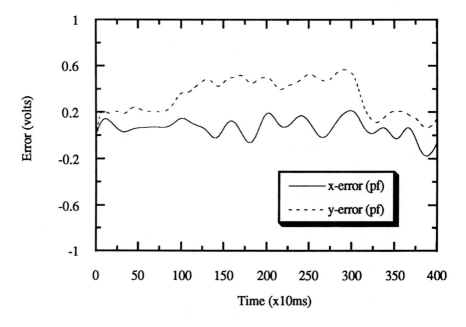

Figure 7-9. Tracking with non-fuzzy proportional feedback.

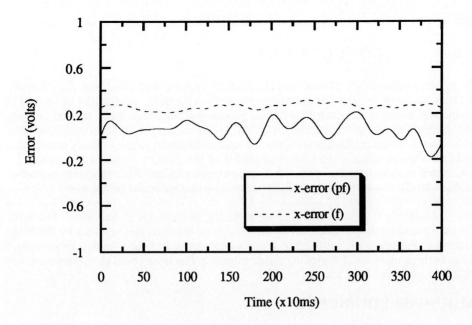

Figure 7-10. Tracking with non-fuzzy proportional feedback and fuzzy feedback.

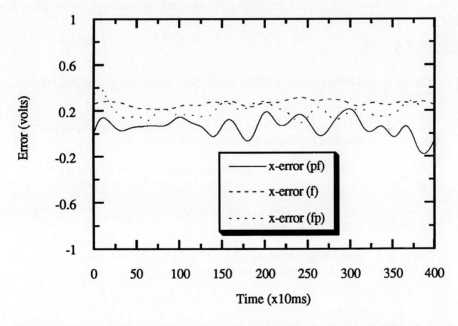

Figure 7-11. Tracking with proportional feedback, fuzzy feedback, and pruned fuzzy feedback.

7.8   CONCLUSIONS

By implementing fuzzy control in this tracking system, overshoot was significantly reduced and limit cycling was virtually eliminated. However, justification of the added complexity of the micro-controller cannot be accomplished until a comparison is made with conventional proportional-plus-integral (PI) control and proportional-plus-derivative (PD) control. These results simply show the ease of implementation of fuzzy control to a problem without using a mathematical model of the system. Future study should be considered in modeling the system and implementing PI and PD controllers to further establish or dispute the viability of fuzzy control for this particular problem.

Additional possibilities include reducing the amount of hardware. The error electronics and nonlinear motor driver circuit can be removed and replaced by the fuzzy controller. Further work can lead to a more optimum set of membership function and rules. Perhaps methods of designing the membership functions could be explored such as techniques described by Wang and Mendel.

ACKNOWLEDGMENTS

Special thanks to Renee Temple who helped me through the tough times by teaching me C, to Denis Barak for help with A/D and D/A processes, and to the Chemical and Laser Sciences Division at Los Alamos who loaned me some equipment that made this project possible. And of course to Dr. Mo Jamshidi, who gave me motivation to accomplish this goal.

REFERENCES

[1]   Klir, G. J., and Folger, T. A. (1988). *Fuzzy Sets, Uncertainty, and Information*, Prentice-Hall.

[2]   Kosko, B. (1992). *Neural Networks and Fuzzy Systems: A Dynamical Systems Approach To Machine Intelligence*, Prentice-Hall.

[3]   Phillips, C. L., and Harbor, R. D. (1988). *Feedback Control Systems*, Prentice-Hall.

[4]   Wang, L. Mendel, J. M. Generating Fuzzy Rules From Numerical Data, With Applications, USC-SIPI Report #169

[5]   Zadeh, L. A. "Knowledge Representation in Fuzzy Logic," *IEEE Transactions on Knowledge and Data Engineering*, Vol. 1, No. 1, March 1989.

Chap. 7   Fuzzy Laser Beam Alignment 161

APPENDIX A

```
/*
      Laser Beam Alignment Fuzzy Inference Engine
      Written using Togai InfraLogic Fuzzy C Software
*/

PROJECT Alignment

   /* Error of the spot measured in 0.039 volts.
    * The universe of discourse is [-5.0, 5.0] volts.
    */

   VAR Error
      TYPE signed byte           /* C type of 'signed char'   */
      MIN -128                   /* universe of discourse min */
      MAX 127                    /* universe of discourse max */

      /* Membership functions for Error (ZE, PS, NS, PM, NM). */

      MEMBER ZE
         POINTS -20 0 0 1 20 0
      END

      MEMBER PS
         POINTS 3 0 20 1 127 0
      END

      MEMBER NS
         POINTS -128 0 -20 1 -3 0
      END

      MEMBER PM
         POINTS 20 0 100 1 127 1
      END

      MEMBER NM
         POINTS -128 1 -100 1 -20 0
      END

   END /* end of Error definition */

   /* dError, the velocity of the spot (or detector),
    * measured in volts.  The universe of discourse
    * [-2,2] volts/second.
```

```
        */
    VAR dError
        TYPE signed byte            /* C type of 'signed char'  */
        MIN -100                    /* universe of discourse min
*/
        MAX 100                     /* universe of discourse max
*/

        /* Membership functions for dError (ZE, PS, NS). */
```

APPENDIX A

```
        MEMBER ZE
            POINTS -30 0 0 1 30 0
        END

        MEMBER PS
            POINTS 0 0 30 1 100 1
        END

        MEMBER NS
            POINTS -100 1 -30 1 0 0
        END
    END /* end of dError definition */

/* Armature voltage of motor in .078 volt increments
 * The universe of discourse is [-128,127] *.078 volts.
 */

    VAR Speed
        TYPE signed byte            /* C type of 'signed char'  */
        MIN -128                    /* universe of discourse min */
        MAX 127                     /* universe of discourse max
*/

        /* Membership functions for Speed (ZE, PS, NS, PM, NM). */

        MEMBER ZE
            POINTS -50 0 0 1 50 0
        END

        MEMBER PS
            POINTS 5 1 50 1 127 0
        END
```

Chap. 7   Fuzzy Laser Beam Alignment

```
    MEMBER NS
        POINTS -128 0 -50 1 -5 0
    END

    MEMBER PM
        POINTS 50 0 125 1 127 1
    END

    MEMBER NM
        POINTS -128 1 -125 1 -50 0
    END

END /* end of Speed definition */

/* Rules for response */

FUZZY Alignment_rules

    RULE Rule1
        IF Error IS PM AND dError IS ZE THEN
            Speed IS NM
    END

    RULE Rule2
        IF Error IS PS AND dError IS PS THEN
            Speed IS NM
    END

    RULE Rule3
        IF Error IS PS AND dError IS NS THEN
            Speed IS ZE
    END

    RULE Rule4
        IF Error IS NM AND dError IS ZE THEN
            Speed IS PM
    END

    RULE Rule5
        IF Error IS NS AND dError IS NS THEN
            Speed IS PM
    END
```

RULE Rule6
    IF Error IS NS AND dError IS PS THEN
        Speed IS ZE
END

RULE Rule7
    IF Error IS ZE AND dError IS ZE THEN
        Speed IS ZE
END

RULE Rule8
    IF Error IS ZE AND dError IS NS THEN
        Speed IS PS
END

RULE Rule9
    IF Error IS ZE AND dError IS PS THEN
        Speed IS NS
END

RULE Rule10
    IF Error IS NS AND dError IS ZE THEN
        Speed IS PS
END

RULE Rule11
    IF Error IS PS AND dError IS ZE THEN
        Speed IS NS
END

RULE Rule12
    IF Error IS NM AND dError IS NS THEN
        Speed IS PM
END

RULE Rule13
    IF Error IS NM AND dError IS PS THEN
        Speed IS PS
END

RULE Rule14
    IF Error IS PM AND dError IS NS THEN
        Speed IS NS
END

Chap. 7   Fuzzy Laser Beam Alignment

```
    RULE Rule15
        IF Error IS PM AND dError IS PS THEN
            Speed IS NM
    END

  END /* end of knowledge base */

/* The following three CONNECT Objects specify that Error
 * and dError are inputs to the Alignment_rules knowledge base
 * and Speed output from Alignment_rules.
 */

    CONNECT
        FROM Error
        TO Alignment_rules
    END

    CONNECT
        FROM dError
        TO Alignment_rules
    END

    CONNECT
        FROM Alignment_rules
        TO Speed
    END

END
```

APPENDIX B

```
/*
        address.h

    Defined constants representing the addresses and masking
    inputs for the A/D and D/A converters.
*/

#include <stdlib.h>
#include <stdio.h>
#include <tilcomp.h>

#define atod_strt 0x0200        /* A/D base address. */
#define atod_data 0x0201        /* A/D data address (base + 1). */
#define atod_ctrl 0x0202        /* A/D control address (base + 2). */
#define atod_stat 0x0203        /* A/D status address (base + 3). */

#define dtoa0_loc 0x0208        /* D/A channel 0 address. */
#define dtoa1_loc 0x020a        /* D/A channel 1 address. */

#define disable_5_ints 0x05     /* Disable A/D channel 5 ext. interupts. */
#define disable_7_ints 0x07     /* Disable A/D channel 7 ext. interupts. */
#define masking_eofc 0x80       /* Status check of end of data conversion. */
```

APPENDIX C

```c
/*
    Fuzzy Laser Beam Tracker driver program
    Richard Marchbanks
    Los Alamos National Laboratory
*/

#include "address.h"
#include "stdio.h"

main()
{
    /* Global Data Variables. */
    char status;
    int err_x,err_y;
    int dxerr,dyerr;
    int xerr, yerr;
    int xerr_old, yerr_old;
    int Error, dError, Speed;
    int i, j, s;

    /* Data Storage. */
    float x_data[500];          /* x sampled data */
    float y_data[500];          /* ysampled data */
    char datafile[15];          /*storage of sampled data */
    FILE *fp;

    /* Initialization. */
    i = 0;
    j = 0;
    s = 1;                      /* sample rate */
    dxerr = 0;
    dyerr = 0;
    xerr_old = 0;
    yerr_old = 0;

    printf("\n\nFUZZY BEAM ALIGNMENT");

    printf("\n\nPlease enter filename:");
    scanf("%s", datafile);

    printf("\nProgram Executing: Press a key to quit.");

    /* While the keyboard is not pressed, execute program. */
```

```c
while(kbhit() == 0)
{
        /* X-ERROR */

        /* Disable interrupt. */
        outportb(atod_ctrl, disable_5_ints);

        /* start a/d conversion */
        outportb(atod_strt,1);

        /* wait until conversion complete */
        while( (status=inportb(atod_stat) & masking_eofc) != 0 );

        /* read the data */
        err_x = inportb(atod_data);

        /* Y-ERROR */

        /* Disable interrupt. */
        outportb(atod_ctrl, disable_7_ints);

        /* start a/d conversion */
        outportb(atod_strt,1);

        /* wait until conversion complete */
        while( (status=inportb(atod_stat) & masking_eofc) != 0 );

         /* read the data */
         err_y = inportb(atod_data);

        /* Write data to array every sth data value. */
        /* Do this for 500 data values and stop saving them*/

        if ((i == s - 1 ) && (j <= 499))          /* sample every s time */
        {
                x_data[j] = (err_x - 128) * .039;
                y_data[j] = (err_y - 128) * .039;
                i = 0;
                j++;
        }
        else
                i++;
```

Chap. 7 Fuzzy Laser Beam Alignment

/* Put the data in signed byte form form for fuzzy controller*/

```
xerr = err_x - 128;
yerr = err_y - 128;

 /* Find delta errors*/

dxerr = xerr - xerr_old;
dyerr = yerr - yerr_old;

/* Update old errors */

xerr_old = xerr;
yerr_old = yerr;

/* Call fuzzy rules */

Alignment(xerr, dxerr*2, &Speed);

err_x = Speed+128;

Alignment(yerr, dyerr*2, &Speed);

 err_y = Speed+128;

/* Write to channel 0 D/A */
err_x = (err_x << 8);
outport(dtoa0_loc, err_x );

/* Write to channel 1 D/A */
err_y = (err_y << 8);
outport(dtoa1_loc, err_y );

}

printf("\n\nProgram Terminated:");
printf("\n\nWriting data to file: %s", datafile);

fp = fopen(datafile, "w");

fprintf(fp, "FUZZY BEAM ALIGNMENT DATA\n\n");
fprintf(fp, "X-ERROR       Y-ERROR\n");

for (i = 0; i <= 499; i++)
```

```
            {
                fprintf(fp, "%f      %f\n", x_data[i], y_data[i]);
            }

        fclose(fp);

        printf("\n\nFile Transfer Complete.\n\n");
}
/* end of driver program */
```

8

INTRODUCTION OF FUZZY SETS IN MANUFACTURING PLANNING

Hädie Fotouhie
University of New Mexico

The goal of this study is to demonstrate a methodology that can enhance decision credibility in the area of manufacturing management. In the following hypothetical case, modification of a production assembly line is investigated. The model would be non-traditional by relying not only on probabilistic methods but also by introducing fuzzy measures into the decision space, that based on an objective function, orders the outcomes according to their desirability. In such manner, decisions can be based on long-term interests by incorporating factors other than those motivated simply by short term and often potentially misleading economical aspects. Such procedure would be particularly effective for international trade, in a new manufacturing world where complexities may arise from, for example, environmental considerations in a way that they can eventually jeopardize the whole venture or investment. Let us assume that a major U.S. car manufacturer is planning to set up an assembly line for one of its models in a foreign country. The product design division has recommended addition of one workstation to the present line in order to make adjustment to the colder and more humid climatic condition in that country, by attachment of a specific anti-corrosion protective shield to the roof and side panels. Such modification is costly, and prior to making the decision the management is interested in knowing how such change will impact the production, what are the risks involved and how will such adjustment pay off within a period of 10 years. The following study provides the management with the information that facilitates the task of decision making through a combination of probabilistic and fuzzy measures.

8.1   INTRODUCTION

The knowledge elicitation technique seems to have a deterministic effect on how fuzzification shapes the decision. However the fuzziness is introduced in the model when decision making under conditions of uncertainty modeled by probability theory cannot be responsive in all aspects. Therefore at such point, fuzziness is entered based on fuzzy information as an attempt to deal with vagueness resulting from imprecise determination of preferences, constraints and even goals. The factor of vagueness could be larger in the case of lack of direct access to the information (in our case, for example, the cultural and historical background of the consumers in the foreign country should ideally be accommodated in the fuzzy data, to more effectively cope with uncertainties). In this study a two-fold approach is employed. The first one concerned with the effect of one additional workstation on production rate, uses the available knowledge on the outcome space and probability distribution related to the performance of each assembly workstation and thus relies on probability measures. The second one links the decision to the fuzzy set theory in order to respond to the conditions of uncertainty where there is no knowledge on the probabilities of the outcomes. To address the uncertainties, one instead has to rely on a range of fuzzy measures in areas like, for example, selecting the most appropriate choice of machine to enter the assembly line while meeting all the requirements, set to conform to a long term strategy.

8.2   THEORY AND FORMULATION

8.2.1 EFFECT OF THE ADDITIONAL WORKSTATION ON PRODUCTION RATE

Like automated flow lines, an automated assembly line consists of several machines or workstations which are linked together by work-handling devices that transport partially assembled parts between stations. The transfer of work parts occurs automatically and the workstations carry out their specialized functions by adding or joining a component in some fashion to an existing assembly. The process steps are performed sequentially as the part moves from one station to the next. The existing assembly consists of a base part plus the components assembled to it at previous stations. When the feed mechanism and the assembly workhead attempt to join a component that is not consistently oriented or is not uniform in size and shape, the station can jam. It can result in shutdown of the entire line until the fault is corrected. The more the number of stations the higher will be the risk of shutdown. Defects occur with a certain fraction defective rate q. Thus in operation of an assembly line, q can be considered as the probability that the next component is defective. Such components being fed into the workstations might or

Chap. 8 Fuzzy Set in Manufacturing Planning

might not cause the station to jam. Based on Groover's approach, if the defective component causes a jam at station i,

$$p_i = m_i q_i \qquad i = 1, 2, \ldots, n \tag{1}$$

shows the probability that a part will jam at station i of an n-station assembly line, where m_i refers to the probability that a defect will cause the ith station to stop. Also note that q_i means the value of q may be different for different stations, so does m_i. The second possible event is when a component is defective but does not cause a station jam. This outcome has the probability

$$(1 - m_i) q_i \tag{2}$$

and finally the third possibility is when component is not defective, with probability

$$1 - q_i \tag{3}$$

It is clear that

$$m_i q_i + (1 - m_i) q_i + (1 - q_i) = 1$$
$$i = 1, 2, \ldots, n \qquad n = \text{\# of stations on the assembly line} \tag{4}$$

The complete distribution of possible outcomes that can occur on an n-station assembly line is the product

$$\prod_{i=1}^{n} [m_i q_i + (1 - m_i) q_i + (1 - q_i)] = 1 \tag{5}$$

Two of the above mentioned three events presented by $m_i q_i$ and $1 - q_i$ result in addition of good components at a given station, thus

$$P_{ap} = \prod_{i=1}^{n} (1 - q_i + m_i q_i) \tag{6}$$

yields the proportion of acceptable products coming off the line.

The frequency of downtime occurrences due to station jams, per cycle is:

$$F = \sum_{i=1}^{n} m_i q_i \tag{7}$$

The average production time per assembly is

$$T_p = T_c + \sum_{i=1}^{n} m_i q_i T_d \qquad (8)$$

where T_c is ideal cycle time and T_d is average downtime per occurrence.

The mean production rate $R_p = \dfrac{1}{T_p}$ in general, however if we want to only take into account the output with no defects,

$$R_{ap} = \frac{\prod_{i=1}^{n}(1 - q_i + m_i q_i)}{T_p} = \frac{P_{ap}}{T_p} \qquad (9)$$

Let in this study $n0=\#$ of stations in the current line set-up=6 and $n_{proposed}=7$. Also assume $T_c=48$ sec., $q=0.01$=fraction defect rate, same for all stations, $m=0.5$ for all the stations and $T_d=20$ min.

For the current line (U.S. version)

$$P_{ap} = [1 - 0.001 + (0.5)(0.01)]^6 = 0.97$$
$$T_p = 0.8 + 6(0.5)(0.01)(20) = 1.4 \min$$
$$(R_{ap})_{o_{currentline}} = \frac{0.97}{1.4} \cong 0.7 \frac{unit}{min} = 42 \frac{units}{hr}$$

In the above calculation, by changing from $n=6$ to $n=7$ due to adding one more station, we will get:

$$(R_{ap})_1 = \text{mean production rate for the proposed line} = \frac{0.965}{1.5} = 39 \frac{units}{hr}$$

that reflects a decrease in production due to the added probability of downtime occurrence. Therefore the additional station will decrease the production by 8640 units/year (based on 360 working days per year, 8 hours per day - one shift).

8.2.2 FUZZINESS INTRODUCED IN SELECTION OF THE ASSEMBLY MACHINE TO BE ATTACHED TO THE PRESENT LINE

Fuzziness can be introduced at a stage in the model where it can be best responsive to the uncertainties that in a way influence the decision. In our case all uncertainties regarding the assembly machine can be handled by fuzzy sets. Based on Bellman and Zadeh, a fuzzy

Chap. 8 Fuzzy Set in Manufacturing Planning

model of decision to accommodate certain constraints C and goals G is suggested. Fuzzy sets C and G project our state of knowledge or preference in a vague form and the membership functions order the outcomes according to the preferability. Both the fuzzy constraints and fuzzy goals can be characterized by membership functions:

$$\mu_{\tilde{C}}: X \to [0,1]$$
$$\mu_{\tilde{G}}: X \to [0,1] \tag{10}$$

X being the universe of alternative interests in a very general sense. However it is possible to consider $\mu_{\tilde{C}}$ on universe X and $\mu_{\tilde{G}}$ on universe Y, where Y is the universe of outcomes.

In this case a mapping function f can be defined as:

$$f: X \to Y \tag{11}$$

Such that a fuzzy goal G defined on set Y induces a corresponding fuzzy goal G' on set X,

$$\mu_{\tilde{G}'}(x) = \mu_{\tilde{G}}[f(x)] \tag{12}$$

A fuzzy decision D may be defined as the choice that satisfies both the constraints C and the goal G,

$$\tilde{D} = \tilde{G} \cap \tilde{C} \tag{13}$$

which in terms of the membership function, using fuzzy set intersection

$$\mu_{\tilde{D}}(x) = \min\left[\mu_{\tilde{G}}(x), \mu_{\tilde{C}}(x)\right] \tag{14}$$

For this case let us consider five different assembly machine options for the task of attaching protective shields to the car body panels, using robotic manipulators at the following costs (in dollars):

$$f(A') = 710,000, f(B') = 326,000, f(C') = 101,200,$$
$$f(D') = 510,000, f(E') = 932,200$$

Our goal is to choose the machine that has the best price given the following constraints defined as fuzzy sets $\tilde{C}_1$ through $\tilde{C}_4$ using Zadeh's shorthand.

C_1 concerns the anti-corrosion quality control factor and it changes for different assembly machines depending on their individual procedures of combined chemical treatment and multiple attachment routines

$$\underset{\sim}{C_1} = \frac{0.70}{A'} + \frac{0.45}{B'} + \frac{0.78}{C'} + \frac{0.61}{D'} + \frac{0.75}{E'} \tag{15}$$

C_2 incorporates pollution parameter resulting from operation of each machine

$$\underset{\sim}{C_2} = \frac{0.71}{A'} + \frac{0.49}{B'} + \frac{0.72}{C'} + \frac{0.57}{D'} + \frac{0.73}{E'} \tag{16}$$

C_3 signifies the reliability and ease of maintenance for each individual machine

$$\underset{\sim}{C_3} = \frac{0.75}{A'} + \frac{0.51}{B'} + \frac{0.80}{C'} + \frac{0.60}{D'} + \frac{0.74}{E'} \tag{17}$$

C_4 deals with the average lifetime of each machine option

$$\underset{\sim}{C_4} = \frac{0.71}{A'} + \frac{0.52}{B'} + \frac{0.85}{C'} + \frac{0.56}{D'} + \frac{0.80}{E'} \tag{18}$$

Therefore the goal can be framed as selecting the most economical option that yields the highest possible product quality with least amount of adverse environmental impact that offers best possible reliability and ease of maintenance and gives the highest average lifetime.

The fuzzy goal G concerning cost of machines is defined on the universal set X by the following membership function

$$\mu_{\underset{\sim}{G}}(x) = -\left(\frac{x}{2 \times 10^5} - 15.0\right)^2 (0.0025) + \frac{1}{10k} + 1.0, \ x \in X \tag{19}$$

$$for\ 1 \times 10^5 \langle x \langle 15 \times 10^5$$

where k is related to average salvage value of machine after 15 years of service. k is defined for each machine option:

Chap. 8 Fuzzy Set in Manufacturing Planning

Table 8-1

Option	k
A'	1.2
B'	0.9
C'	1.4
D'	1.1
E'	1.5

Such function f on G yields a corresponding goal G' induced on the set of alternative machine options as

$$\underset{\sim}{G'} = \frac{0.75}{A'} + \frac{0.66}{B'} + \frac{0.82}{C'} + \frac{0.70}{D'} + \frac{0.79}{E'} \tag{20}$$

taking the standard fuzzy set intersection on the 5 fuzzy sets, we get to the fuzzy decision set $\underset{\sim}{D}$;

$$\underset{\sim}{D} = \underset{\sim}{G'} \cap \underset{\sim}{C_1} \cap \underset{\sim}{C_2} \cap \underset{\sim}{C_3} \cap \underset{\sim}{C_4} \tag{21}$$

$$= \min\left[\mu_{\underset{\sim}{G'}}(x), \mu_{\underset{\sim}{C_1}}(x), \mu_{\underset{\sim}{C_2}}(x), \mu_{\underset{\sim}{C_3}}(x), \mu_{\underset{\sim}{C_4}}(x)\right] \tag{22}$$

for our case;

$$\underset{\sim}{D} = \frac{0.70}{A'} + \frac{0.45}{B'} + \frac{0.72}{C'} + \frac{0.56}{D'} + \frac{0.73}{E'} \tag{23}$$

Now, taking the maximum of this decision set gives the optimum alternative as the option that seems best to satisfy both our goal and constraints. In this case, machine E would be the choice

$$D = \max\left[\min\left(\mu_{\underset{\sim}{G'}}, \mu_{\underset{\sim}{C_1}}, ..., \mu_{\underset{\sim}{C_4}}\right)\right] = 0.73 \quad \Rightarrow \quad \text{machine } E' \tag{24}$$

8.2.3 IS USING THE CURRENT 6-STATION SET-UP (NO ASSEMBLY LINE MODIFICATION) AN OPTION?

In order to come up with the answer prior to making a final decision, a survey was done by 2 independent sources overseas. The motivation was based on the following: knowing that for $0.70 < C_1 \int$ (rust protection quality factor) < 0.80, as is also the case for the selected machine E' ($C_1 = 0.75$), based on observation (statistics) from certain northeastern states that have some similarity with the European site, in terms of climate, it is decided that annual sales drops will be contained within 2 percent. However it is predicted that if no adjustment is made in the product line (meaning that C_1 is even less than 0.45 which is the case for the lowest machine B'), the sales will plunge somewhere between 2 to 10 percent each year. Therefore the survey aims at three sets A, B and C as 3, 5 and 7 percent, respectively, signifying annual sales drops due to implementation of the U.S. version of the assembly line in the foreign site. The two independent sources have collected evidence by different means and methods and have come up with basic assignments m_1 and m_2. These are the degrees of evidence that support to what extent these 3 claims A, B, or C are true. The result of the survey shows:

Table 8-2

m_2	Focal elements	m_1
0.04	A	0.01
0.15	B	0.08
0.08	C	0.12
0.12	A»B	0.09
0.07	A»C	0.10
0.24	B»C	0.25
0.30	A»B»C	0.35

Using Dempster's rule of combination we proceed to combine the bodies of evidence to obtain joint basic assignments. In such manner by employing evidences from both sources and merging them together, the risk of being misled would be minimized.

In general

$$m_{1,2}(A) = \frac{\sum_{B \cap C = A} m_1(B) m_2(C)}{1 - k} \quad (25)$$

$$k = \sum_{B \cap C = 0} m_1(B) m_2(C) \quad (26)$$

Chap. 8   Fuzzy Set in Manufacturing Planning

$k = m_1(A) \cdot m_2(B) + m_1(A) \cdot m_2(C) + m_1(A) \cdot m_2(B \cup C) + m_1(B) \cdot m_2(A) + m_1(B) \cdot m_2(C) + m_1(B) \cdot m_2(A \cup C) + m_1(C) \cdot m_2(A) + m_1(C) \cdot m_2(B) + m_1(C) \cdot m_2(A \cup B) + m_1(A \cup B) \cdot m_2(C) + m_1(A \cup C) \cdot m_2(B) + m_1(B \cup C) \cdot m_2(A) = 0.089$

$\Rightarrow$ normalization factor $1 - k = 0.91$ \hfill (27)

$$m_{1,2}(A) = \frac{1}{1-k}\begin{bmatrix} m_1(A) \cdot m_2(A) + m_1(A)m_2(A \cup B) + m_1(A)m_2(A \cup C) + m_1(A)m_2(A \cup B \cup C) + m_1(A \cup B)m_2(A) + m_1(A \cup B)m_2(A \cup C) + m_1(A \cup C) \\ m_2(A) + m_1(A \cup C)m_2(A \cup B) + m_1(A \cup B \cup C)m_2(A) \end{bmatrix} = 0.05$$

(28)

$$m_{1,2}(B) = \frac{1}{1-k}\begin{bmatrix} m_1(B) \cdot m_2(B) + m_1(B)m_2(A \cup B) + m_1(B)m_2(B \cup C) + m_1(B)m_2(A \cup B \cup C) + m_1(A \cup B)m_2(B) + m_1(A \cup B)m_2(B \cup C) + m_1(B \cup C) \\ m_2(B) + m_1(B \cup C)m_2(A \cup B) + m_1(A \cup B \cup C)m_2(B) \end{bmatrix} = 0.24$$

(29)

$$m_{1,2}(C) = \frac{1}{1-k}\begin{bmatrix} m_1(C) \cdot m_2(C) + m_1(C)m_2(A \cup C) + m_1(C)m_2(B \cup C) + m_1(C)m_2(A \cup B \cup C) + m_1(A \cup C)m_2(C) + m_1(A \cup C)m_2(B \cup C) + m_1(B \cup C) \\ m_2(C) + m_1(B \cup C)m_2(A \cup C) + m_1(A \cup B \cup C)m_2(C) \end{bmatrix} = 0.19$$

(30)

Therefore the joint bodies of evidence support the claim of <u>five</u> percent per year drop in product sale due to using the present 6-station assembly line configuration and $C_1 < 0.45$.

8.3   RESULTS AND DISCUSSION

a) **The modified line:**
As we obtained earlier in this study, by the line adjustment (n=7), we not only had to spend \$932,200 to add the assembly machine (version E') to the U.S. version of the assembly line, but such decision causes the production to drop from 42 units/hr to 39 units/hr, or 8640 units less will be produced each year. Let the gross value of each unit be \$5000. Also consider the estimated 2 percent annual sales drop for $0.7 < C1 < 0.8$ as the machine E' dictates. 39 units/hr fi 39 X 8 X 360 = 112,320 units/year = 112,320 X \$5000 = \$5.61 X 10^8. Considering the initial cost of machine E', as \$932,200, we obtain 5.61 X 10^8 - 932,200 = \$5.6 X 10^8 as the sales value for the first year.

After five years it will drop to 5.6 X 10^8 X [1-(5 X $\frac{2}{100}$)]=\$5.04 X 10^8 and after 10 years, to \$4.5 X 10^8.

b)   The 6-station present line:
42 units/hr = 42 X 8 X 360 = 120,960 units/year = \$6.08 X 10^8 will be the sales value for the first year. After 5 years it will drop to 6.08 X 10^8 X [1-(5 X $\frac{5}{100}$)] = \$4.56 X 10^8 and within a period of 10 years to \$3.4 X 10^8.

8.4   CONCLUSION

The complexity of a decision-making process usually arises from the informational uncertainty within the determinants that cannot be handled by probabilistic methods. A decision may not be responsive in every aspect until it takes into account all the pertinent information that in different ways intersect it, especially in cases where a long-term strategy is to be adopted. Application of the concept of fuzzy set theory will give the tool to cope with imprecision or generalities, mathematically through the membership functions. By means of departure from binary logic, the decision maker will be able to incorporate in the decision all the sensitive issues that otherwise will deteriorate the quality of the decision in a matter of time.

REFERENCES

Baecher, G. B. "Professional judgment and prior probabilities in engineering risk assessment." Fourth International Conference on Applications of Statistics and Probability in Soil and Structural Engineering, University di Firenze, Italy, 1983.

Bel, G., Bensana, E., Dubois, D., Koning, J. L. "Handling fuzzy priority rules in a job-shop scheduling system." IFSA proceedings, 1989.

Groover, M. P. *Automation, Production Systems and Computer Integrated Manufacturing.* Prentice Hall, 1987.

Klir, G. J. and Folger, T. A. *Fuzzy Sets, Uncertainty and Information.* Prentice Hall, 1988.

Zimmermann, H. J. "The use of fuzzy LP and approximate reasoning in production scheduling." IFSA proceedings, 1989.

9

A COMPARISON OF CRISP AND FUZZY LOGIC METHODS FOR SCREENING ENHANCED OIL RECOVERY TECHNIQUES

W. J. Parkinson
and K. H. Duerre
Los Alamos National Laboratory

9.1 INTRODUCTION

Reasons for studying enhanced oil recovery (EOR) techniques are summarized in a 1986 paper by Stosur [1]. When his paper was published, only 27% of all the oil discovered in the United States had been produced. Under current economic conditions, only about 6% more will be produced using existing technology. The remaining 67% is a target for EOR. Currently, about 6% of our daily oil production comes from EOR. Even in these times of reduced concern of an energy crisis, these numbers indicate that the study of EOR processes can be rewarding because of the potentially high payoffs.

Because, in general, EOR processes are expensive, it is necessary for engineers to pick the best recovery method for the reservoir in question to optimize profits or to make any profits at all. The screening methods are expensive and typically involve many steps, one of which is to consult the technical screening guide; this screening step is the subject of this chapter. Screening guides consist of tables or charts that list the rules of thumb for picking a proper EOR technique as a function of reservoir and crude oil properties. Once a candidate EOR techniques is determined, further laboratory flow studies are often required. Data obtained from these studies are then used to demonstrate the viability of the selected technique. Throughout the screening process, economic evaluations are carried out .

In this chapter, we present two expert systems for screening EOR processes. In the first, we developed a crisp, rule-based assistant, which replaces the previously published screening guides. These guides are based on tables and graphs designed for hand calculations. The expert assistant provides essentially the same information as the table and graph method, but is more comprehensive and easier to use than the screening guides. The second, fuzzy expert assistant was then developed to eliminate some of the weaknesses observed in the first expert system. At the end of the test session, both of these expert assistants provide users with a ranked list of potential techniques. This is difficult to do using the tables. With both expert systems, the user must enter oil gravity, viscosity, composition, formation salinity, formation type, oil saturation, thickness, permeability, depth, temperature, and porosity. Although the final choice of technique will be based upon economics, the first screening step is quite important because the screening process itself is expensive and because of the absolute necessity of choosing the most economically optimum EOR technique.

9.2   THE EOR SCREENING PROBLEM

For this study, EOR is defined as any technique that increases production beyond water flooding or gas recycling. This usually involves the injection of an EOR fluid. Both of the expert systems discussed here are rule based and both rely mainly on the work of Taber and Martin [2] and Goodlet et al. [3,4] for their rules.

EOR techniques can be divided into four general categories: thermal, gas injection, chemical flooding, and microbial. Thermal techniques are then subdivided into *in situ* combustion and steam flooding, which require reservoirs with fairly high permeability. Steam flooding has, traditionally, been the most used EOR method. It was previously applied only to relatively shallow reservoirs containing viscous oils. In this application, screening criteria are changing because the improved equipment allows economic operations on deeper formations. New studies show that, in addition to their effect on viscosity and density, steam temperatures also affect other reservoir rock and fluid properties. Thus, reservoirs previously not considered as candidates for steam flooding are being reevaluated. The expert system format is a good one to use here because we can easily change the program as the knowledge of a technology changes. Gas injection techniques, however, are at the opposite extreme from steam flooding. They are divided into hydrocarbon, nitrogen and flue gas, and carbon dioxide. These techniques tend to work best in deep reservoirs containing light oils. Chemical flooding techniques are

Chap. 9 Fuzzy Logic for Enhanced Oil Recovery

divided into polymer, surfactant/polymer, and alkaline recovery techniques. Microbial techniques are new, and primarily experimental, at this time. The microbial category is not subdivided. Figure 9-1 shows all four of these categories and their associated EOR methods as the search tree for both expert assistants.

The comment is often heard, "We have excellent papers on this subject with graphs and tables and information to help us solve the problem. Why do we need an expert system?" The response is that an expert system is not absolutely necessary, but the problem can be solved more quickly, and often better, with the expert system. Table I, taken directly from Ref. 2, is a matrix of eight EOR techniques and nine EOR criteria.

Theoretically, if the values of the EOR criteria for the reservoir in question are known, engineers can pick some candidate processes from Table I, even without having much knowledge about EOR.

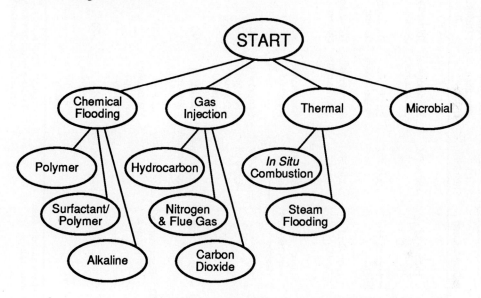

Figure 9-1. Search tree for the expert assistants.

The following simple examples show some of the problems with this argument. For Example 1, the following EOR criteria are used with Table I:

Example 1

(1) Gravity = 18 degrees API
(2) Viscosity = 500 cp
(3) Composition = high percent of $C_4 - C_7$
(4) Oil saturation = 50%
(6) Payzone thickness = 35 ft
(7) Average permeability = 1000 md.
(8) Well depth = 2000 ft
(9) Temperature = 110°F.

TABLE I. Summary for screening criteria for enhanced recovery methods[c]

	Oil Properties				Reservoir Characteristics				
	Gravity °API	Viscosity (cp)	Composition	Oil Saturation	Formation Type	Net Thickness (ft)	Average Permeability (md)	Depth (ft)	Temperature (°F)
Gas Injection Methods									
Hydrocarbon	>35	<10	High % of C_2–C_7	>30% PV	Sandstone or Carbonate	Thin unless dipping	NC	<2000 (LPG) to <5000 HP gas	NC
Nitrogen & Flue Gas	>24 >35 for N_2	<10	High % of C_1–C_7	>30% PV	Sandstone or Carbonate	Thin unless dipping	NC	>4500	NC
Carbon Dioxide	>26	<15	High % of C_5–C_{12}	>30% PV	Sandstone or Carbonate	Thin unless dipping	NC	>2000	NC
Chemical Flooding									
Surfactant/ Polymer	>25	<30	Light intermediates desired	>30% PV	Sandstone preferred	>10	>20	<8000	<175
Polymer	>25	<150	NC	>10% PV Mobile oil	Sandstone Preferred Carbonate possible	NC	>10 (normally)	<9000	<200
Alkaline	13-35	<200	Some organic acids	Above water-flood residual	Sandstone preferred	NC	>20	<9000	<200
Thermal									
Combustion	<40 (10-25 normally)	<1000	Some asphaltic components	>40-50% PV	Sand or Sandstone with high porosity	>10	>100[a]	>150 preferred	>150 preferred
Steam Flooding	<25	>20	NC	>40-50% PV	Sand or Sandstone with high porosity	>20	>200[b]	300-5000	NC

NC = not critical
[a] Transmissibility >20 md ft/cp
[b] Transmissibility >100 md ft/cp
[c] From reference 2

Copyright 1983, Society of Petroleum Engineers, Taber, J.J. and Martin, F.D.; "Technical Screening Guides for the Enhanced Recovery of Oil," paper SPE 12069 presented at the 1983 Annual Technical Conference, San Francisco, CA, October 5-8.

Chap. 9   Fuzzy Logic for Enhanced Oil Recovery

If we search the table, starting at the top, and move left-to-right before moving down a row, we are using backward-chaining or a goal-driven method. That is, we first assume a solution (e.g. hydrocarbon gas-injection), then check the data either to verify or to disprove that assumption. On the other hand, a data-driven or forward-chaining approach would begin the search in the upper left-hand corner of the table and would move down, row by row, to the bottom before moving to the next column. That is, the search would start with the datum value for the oil gravity and would check that value against every EOR method before moving on to the other data. In this example, we use backward-chaining to find that steam flooding is the only good method to use for this example. The results of this search are shown in Figure 9-2. *In situ* combustion techniques might also work.

The preceding situation, is not ideal because there is only one candidate for the next screening step, and this candidate could be eliminated, for other reasons, in a later screening step; then there would be no candidate recovery methods for this example. Having a reservoir that is not recommended for EOR is certainly legitimate, but we shouldn't eliminate the possibility of EOR because of too little knowledge. By changing the previous example slightly, we can have the opposite problem, as shown in Example 2, which has the following values for the EOR criteria:

Example 2
- (1) Gravity = 35 degrees API
- (2) Viscosity = 5 cp
- (3) Composition high percent of C_4 - C_7 and some organic acids
- (4) Oil saturation 50%
- (5) Formation type = sandstone
- (6) Payzone thickness = 10 ft
- (7) Average permeability = 1000 md.
- (8) Well depth = 5000 ft
- (9) Temperature = 150°F.

By searching Table I, again with a backward-chaining technique, we obtain the results shown in Figure 9-3. This time only the steam flooding EOR method has been eliminated. This takes us to the second step with, possibly, too many candidates.

This is not a criticism of Ref. 2 or of tables like Table I. It is merely an effort to point out that in order to do a good first screening step, we will often need more information than is available in these tables. Much of this needed information is available in Refs. 2-4. References 3 and 4 include a tables similar to Table I. Table II contains all of the material from Table I, as well as some of the information from the table in Ref. 4, including the microbial drive EOR method. The additional information improves the results of our search, but is still insufficient. We need information that will tell us what the impact of a reservoir temperature of 110°F will be when, as listed in Tables I and II, a temperature of greater than 150°F is preferred. We also need information that will help us rank two or more methods when the methods fall within the acceptable range. That is, we need a ranked list of methods. A nonexpert can obtain a ranked list by reading the papers, and, possibly, by undertaking a short literature search, in addition to using Table I or II.

TABLE II. Summary for screening criteria for enhanced recovery methods[e]

	Oil Properties				Reservoir Characteristics						
	Gravity °API	Viscosity (cp)	Composition	Salinity (ppm)	Oil Saturation	Formation Type	Net Thickness (ft)	Average Permeability (md)	Depth (ft)	Temperature (°F)	Porosity (%)
Gas Injection Methods											
Hydrocarbon	>35	<10	High % of C_2–C_7	NC	>30% PV	Sandstone or Carbonate	Thin unless dipping	NC	>2000 (LPG) to >5000 HP-gas	NC	NC
Nitrogen & Flue Gas	>24 >35 for N_2	<10	High % of C_1–C_7	NC	>30% PV	Sandstone or Carbonate	Thin unless dipping	NC	>4500	NC	NC
Carbon Dioxide	>26	<15	High % of C_5–C_{12}	NC	>30% PV	Sandstone or Carbonate	Thin unless dipping	NC	>2000	NC	NC
Chemical Flooding											
Surfactant/ Polymer	>25	<30	Light intermediates desired	<140,000	>30% PV	Sandstone preferred	>10	>20	<8000	<175	>20
Polymer	>25	<150	NC	<100,000	>10% PV Mobile oil	Sandstone Preferred Carbonate possible	NC	>10 (normally)	<9000	<200	≤20
Alkaline	13-35	<200	Some organic acids	<100,000	Above water-flood residual	Sandstone preferred	NC	>20	<9000	<200	≥20
Thermal											
Combustion	<40 (10-25 normally)	<1000	Some asphaltic components	NC	>40-50% PV	Sand or Sandstone with high porosity	>10	>100[a]	>500	>150 preferred	≥20[c]
Steam Flooding	<25	>20	NC	NC	>40-50% PV	Sand or Sandstone with high porosity	>20	>200[b]	300-5000	NC	≥20[d]
Microbial											
Microbial Drive	>15	*	Absence of toxic conc. of metals, No biocides present	<100,000	NC	Sandstone or Carbonate	NC	>150	<8000	<140	–

NC = not critical
[a]Transmissibility > 20md ft/cp
[b]Transmissibility >100 md ft/cp
[c]Ignore if saturation times porosity >0.08
[d]Ignore if saturation times porosity >0.1
[e]Modified from Refs. 2 and 4

Gas Injection Methods	Gravity	Viscosity	Composition	Oil Saturation	Formation Type	Net Thickness	Average Permeability	Depth	Temperature
Hydrocarbon	no								↑
Nitrogen & Flue Gas	no								↑
Carbon Dioxide	no								↑
Chemical Flooding									
Surfactant/Polymer	no								↑
Polymer	no								↑
Alkaline	yes	no							↑
Thermal									
Combustion	yes	yes	yes	yes	yes	yes	yes	yes	no
Steam Flooding	yes	yes	NC	yes	yes	yes	yes	yes	NC
NC = not critical									

Figure 9-2. Solution to Example Problem 1.

187

Gas Injection Methods	Gravity	Viscosity	Composition	Oil Saturation	Formation Type	Net Thickness	Average Permeability	Depth	Temperature
Hydrocarbon	yes	yes	ok	yes	yes	ok	NC	yes	NC
Nitrogen & Flue Gas	yes	yes	ok	yes	yes	ok	NC	yes	NC
Carbon Dioxide	yes	yes	ok	yes	yes	ok	NC	yes	NC
Chemical Flooding									
Surfactant/Polymer	yes	yes	ok	yes	yes	yes	yes	yes	yes
Polymer	yes	yes	NC	yes	yes	NC	yes	yes	yes
Alkaline	yes	yes	ok	yes	yes	NC	yes	yes	yes
Thermal									
Combustion	yes	yes	ok	yes	yes	yes	yes	yes	NC
Steam Flooding	no								↑

NC = not critical

Figure 9-3. Solution to Example Problem 2.

Chap. 9   Fuzzy Logic for Enhanced Oil Recovery   189

But the time required for this screening step may be far greater than the few minutes required for searching the tables. If the exercise has to be repeated several times or by several different nonexperts, then a small PC-based expert system can be easily justified for the job.

Figures 9-4 to 9-14 demonstrate the basis of a scoring system for the various EOR criteria and for the EOR methods used in a first attempt to solve this problem using a crisp rule-based expert system (see reference 5). Figures 9-5, 9-11, and 9-12 were taken from Ref. 2 and modified. The others were created by studying Ref. 2 through 4 and 6 through 8. Figures 9-4 to 9-14 are bar graphs showing the relative influence that each EOR criterion has on each EOR method. The scoring system is empirical and was designed to add some judgment or expertise to the expert system. A great deal of work went into developing this scoring system

Figures 9-4 to 9-14 demonstrate the basis of a scoring system for the various EOR criteria and for the EOR methods used in a first attempt to solve this problem using a crisp rule-based expert system (see reference 5). Figures 9-5, 9-11, and 9-12 were taken from Ref. 2 and modified. The others were created by studying Ref. 2 through 4 and 6 through 8. Figures 9-4 to 9-14 are bar graphs showing the relative influence that each EOR criterion has on each EOR method. The scoring system is empirical and was designed to add some judgment or expertise to the expert system. A great deal of work went into developing this scoring system

	0	20	40	60	80	100
Hydrocarbon Miscible	poor		good	preferred		
Nitrogen & Flue Gas	poor		*	preferred		
Carbon Dioxide	possible**		fair	good		
Surfactant/Polymer	poor			preferred		
Polymer Flooding	poor			preferred		
Alkaline Flooding	poor***		preferred	fair		
In Situ Combustion	fair		pref.	fair	poor	
Steam Flooding	fair		pref.	poor		
Microbial Drive	poor			good		

* Minimum preferred, 24 for flue gas and 35 for nitrogen.
** Possible immiscible gas displacement.
*** No organic acids are present at this gravity.

Figure 9-4. Oil gravity screening data (°API).

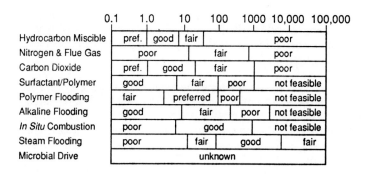

Figure 9-5. Oil viscosity screening data (cp).

	High % C_2-C_7	High % C_1-C_7	High % C_5-C_{12}	Organic Acids	Asphaltic Components
Hydrocarbon Miscible	preferred	good	fair	NC	NC
Nitrogen & Flue Gas	good	preferred	fair	NC	NC
Carbon Dioxide	fair	fair	preferred	NC	NC
Surfactant/Polymer	fair	fair	preferred	NC	NC
Polymer Flooding	NC	NC	NC	NC	NC
Alkaline Flooding	NC	NC	NC	preferred	NC
In Situ Combustion	NC	NC	NC	NC	preferred
Steam Flooding	NC	NC	NC	NC	NC
Microbial Drive	NC	NC	NC	NC	NC

NC = not critical

Figure 9-6. Oil composition screening data.

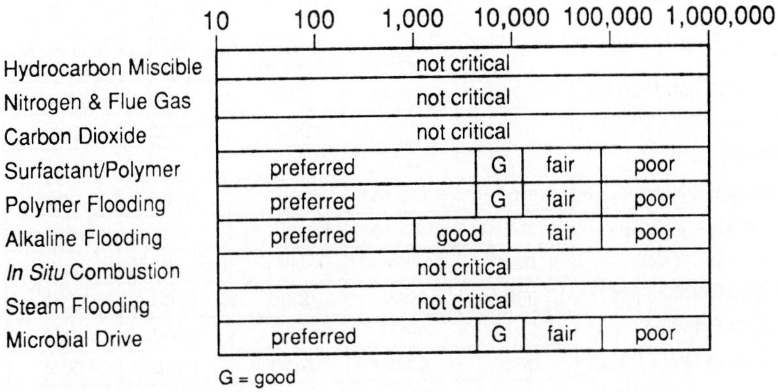

	10	100	1,000	10,000	100,000	1,000,000
Hydrocarbon Miscible	not critical					
Nitrogen & Flue Gas	not critical					
Carbon Dioxide	not critical					
Surfactant/Polymer	preferred			G	fair	poor
Polymer Flooding	preferred			G	fair	poor
Alkaline Flooding	preferred			good	fair	poor
In Situ Combustion	not critical					
Steam Flooding	not critical					
Microbial Drive	preferred			G	fair	poor

G = good

Figure 9-7. Formation salinity screening data (ppm).

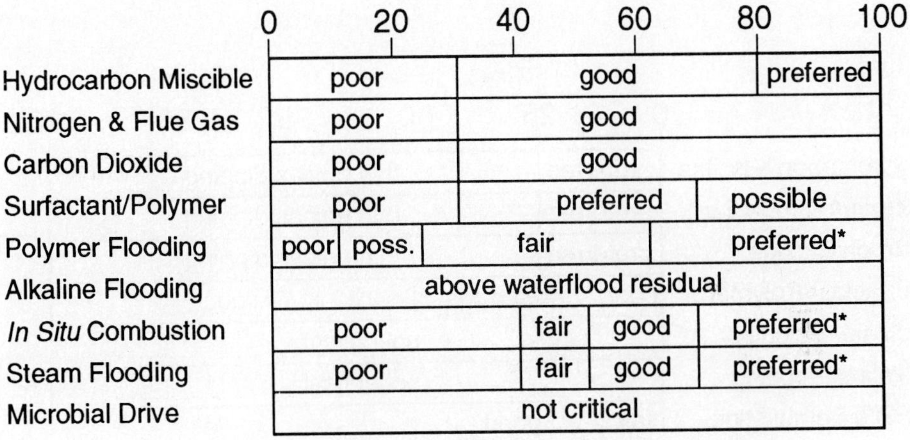

	0	20	40	60	80	100
Hydrocarbon Miscible	poor		good		preferred	
Nitrogen & Flue Gas	poor		good			
Carbon Dioxide	poor		good			
Surfactant/Polymer	poor		preferred			possible
Polymer Flooding	poor	poss.	fair			preferred*
Alkaline Flooding	above waterflood residual					
In Situ Combustion	poor			fair	good	preferred*
Steam Flooding	poor			fair	good	preferred*
Microbial Drive	not critical					

* Preferred status is based on the starting residual oil saturations of successfully producing wells as documented by Ref. 8.

Figure 9-8. Oil saturation screening data (%PV).

	Sand	Homogeneous Sandstone	Heterogeneous Sandstone	Homogeneous Carbonate	Heterogeneous Carbonate
Hydrocarbon Miscible	good	good	poor	good	poor
Nitrogen & Flue Gas	good	good	poor	good	poor
Carbon Dioxide	good	good	poor	good	poor
Surfactant/Polymer	preferred	preferred	poor	good	poor
Polymer Flooding	preferred	preferred	good	fair	poor
Alkaline Flooding	poor	preferred	fair	not feasible	not feasible
In Situ Combustion	good	good	good	good	fair
Steam Flooding	good	good	fair	good	fair
Microbial Drive	good	good	good	good	poor

Figure 9-9. Formation type screening data.

	0–25	25–50	50–75	75–80	80–>100
Hydrocarbon Miscible	preferred	thin unless dipping			
Nitrogen & Flue Gas	preferred	thin unless dipping			
Carbon Dioxide	preferred	thin unless dipping			
Surfactant/Polymer	poor	preferred	good		
Polymer Flooding	not critical				
Alkaline Flooding	not critical				
In Situ Combustion	fair	good	fair		
Steam Flooding	poor	fair	preferred	good	
Microbial Drive	not critical				

Figure 9-10. Net thickness screening data (feet).

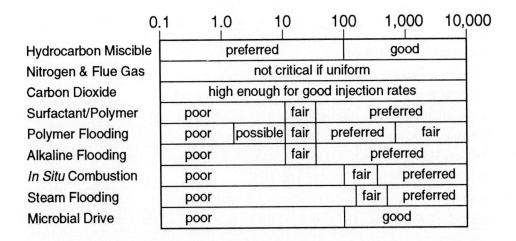

Figure 9-11. Permeability screening data (md).

Figure 9-12. Well-depth screening data (feet).

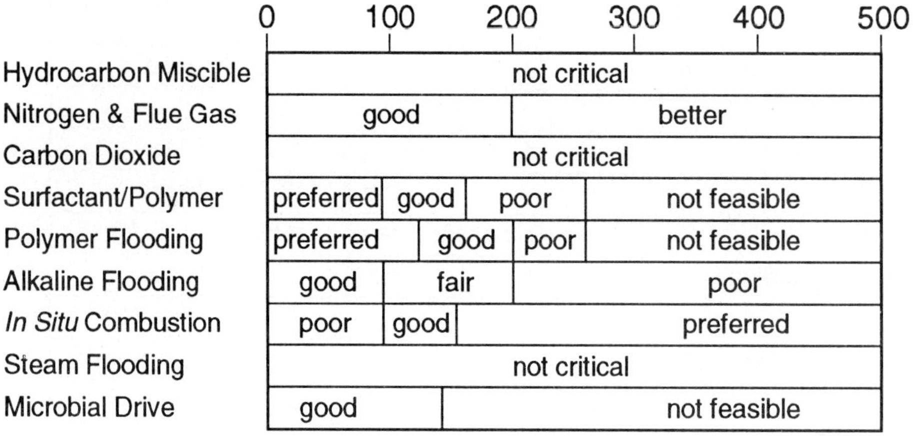

	0	100	200	300	400	500
Hydrocarbon Miscible			not critical			
Nitrogen & Flue Gas		good			better	
Carbon Dioxide			not critical			
Surfactant/Polymer	preferred	good	poor		not feasible	
Polymer Flooding	preferred		good	poor	not feasible	
Alkaline Flooding	good		fair		poor	
In Situ Combustion	poor	good		preferred		
Steam Flooding			not critical			
Microbial Drive		good			not feasible	

Figure 9-13. Formation temperature screening data (°F).

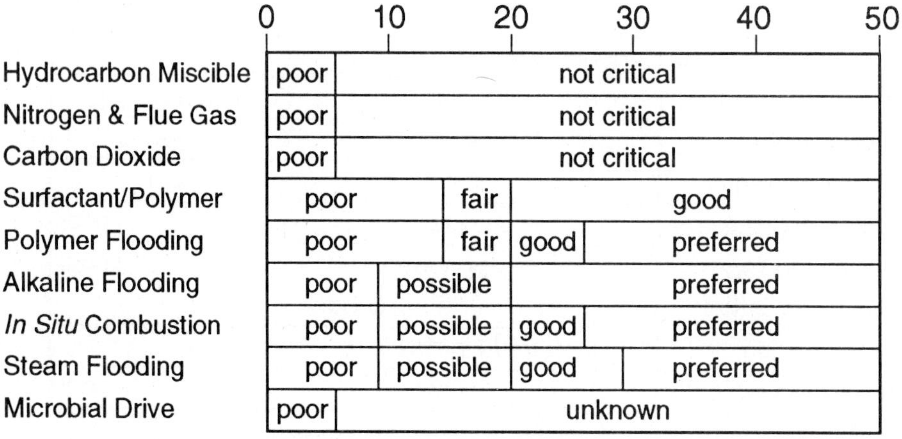

	0	10	20	30	40	50
Hydrocarbon Miscible	poor		not critical			
Nitrogen & Flue Gas	poor		not critical			
Carbon Dioxide	poor		not critical			
Surfactant/Polymer	poor		fair		good	
Polymer Flooding	poor		fair	good	preferred	
Alkaline Flooding	poor	possible		preferred		
In Situ Combustion	poor	possible	good	preferred		
Steam Flooding	poor	possible	good	preferred		
Microbial Drive	poor		unknown			

Figure 9-14. Formation porosity screening data (%).

Chap. 9   Fuzzy Logic for Enhanced Oil Recovery

The system is a significant improvement over the tables because each category is broken into many increments or sets. However, this system is still not adequate because the sets are crisp and they have a membership of either 0 or 1. This works fine for many problems but not for others. For example, in Figure 9-13, the influence of the formation temperature on the microbial drive method is tremendous. With a change of one degree, the choice can go from "Good" to "Not Feasible." This is a change of 60 points. Although there is a temperature above which the microbes die, it is unlikely that the demarcation is that sharp. The crisp scoring system is based on the key words in Figures 9-4 to 9-14, and works like this:

Not feasible	-50	Fair	6
Very poor	-20	Good	10
Poor	0	Not critical	12
Possible	4	Preferred	15

Note that "Not Critical" is a very good situation to have.

For the microbial drive method, the effect of viscosity, and, to a large extent, porosity, is unknown. Until more information is obtained, they are assigned a grade of 6 for an "Unknown," which is the same score as a "Fair."

For an example of the scoring system, observe Figure 9-5 and consider an oil with a viscosity of about 500 centipoise. The hydrocarbon gas injection, surfactant-polymer, and alkaline chemical flood techniques are all "Poor," with scores of zero. The other two gas injection techniques, nitrogen and flue gas and carbon dioxide, are both "Fair," with scores of 6. The polymer flooding technique cannot be used with a viscosity this high, so it gets a score of -50. Each of the thermal techniques is "Good," and each gets a score of 10. The microbial drive method has an "Unknown," so it gets a score of 6.

Some EOR criteria carry more weight than others, and, in some cases, a given criteria may affect one method more than another, which explains why the maximum and minimum scores for each method vary within a given criteria (see Figure 9-4). The variation in oil gravity allows the score of the hydrocarbon miscible gas injection method to range from "Poor" to "Preferred," a point spread of 0 to 15. The same gravity variation allows the score of the carbon dioxide gas injection method to range from "Possible" to "Good," a point spread of 4 to 10. This indicates that oil gravity has a larger influence on the hydrocarbon miscible method than on the carbon dioxide method. Much of the information in Figures 9-4 to 9-14 is based on experience and judgment, and it is influenced by the study of the more than 200 EOR projects listed in Ref. 8. The scoring system used in either expert system can easily be changed by someone with different experience or with new information.

Although scoring system described does quite well in most cases, there are some notable exceptions. These are described in the next two examples. Example 3 has the following values for the EOR criteria for two similar scenarios:

Example 3 - Scenario One
   (1)   Gravity = 23 degrees API
   (2)   Viscosity = 30 cp

(3) Composition = high percent C_5 -C_{12}
(4) Salinity = 101,000 ppm
(5) Oil saturation = 29%
(6) Formation type = sandstone (homogeneous)
(7) Payzone thickness = 26 ft
(8) Average permeability = 24 md
(9) Well depth = 1999 ft
(10) Temperature = 91°F
(11) Porosity = 19%

Scenario Two
(1) Gravity = 24 degrees API
(2) Viscosity = 22 cp
(3) Composition = high percent of C_5 - C_{12}
(4) Salinity = 99,000 ppm
(5) Oil Saturation = 31 %
(6) Formation type = sandstone (homogeneous)
(7) Payzone thickness = 24 ft
(8) Average permeability = 26 md
(9) Well depth = 2001 ft
(10) Temperature = 89°F
(11) Porosity = 21 %

The differences between these two scenarios are hardly measurable. Yet the crisp expert system gives them the following rankings and raw scores:

Scenario One (Rankings)

1-	Polymer flooding	102 points
2-	Alkaline flooding	97 points
3-	*In situ* combustion	93 points
4-	Steam flooding	92 points
5-(tie)	Microbial drive	88 points
6-(tie)	Surfactant/polymer	88 points
7-	Carbon dioxide	85 points
8-	Hydrocarbon miscible	77 points
9-	Nitrogen and flue gas	72 points

Scenario Two (Rankings)

1-	Surfactant/polymer	142 points
2-	Polymer flooding	136 points
3-	Alkaline flooding	127 points
4-	Carbon dioxide	116 points
5-	Nitrogen and flue gas	114 points
6-	Hydrocarbon miscible	104 points
7-	Microbial drive	94 points
8-	Steam flooding	83 points
9-	*In situ* combustion	80 points

Chap. 9   Fuzzy Logic for Enhanced Oil Recovery   **197**

It is easy to see that the rankings of these scenarios are completely different. The scores for the second scenario, except for *in situ* combustion and steam flooding, are much higher than those for the first scenario. (The relevance of the magnitude of these scores is discussed at the end of this section.) A verification of these scores and the reason for the differences are shown in Figures 9-4 to 9-14. These figures show that the scores for many of the EOR methods fall on one side of a crisp boundary in the first scenario and on the other side in the second scenario. The differences are increased because this occurs several times for each method as the expert system searches through the EOR criteria. This example is a worst case. It was set up so that the differences in scores would propagate, rather than cancel, from one criteria to another. But it is realistic in that most measurement techniques are not accurate enough to determine which side of a crisp boundary the data should really be on. The problem is exacerbated by the fact that a small change in the state of an EOR criterion can dramatically influence some EOR methods. For example, Figure 9-4 shows that a small change in the API gravity of an oil can change the potential for surfactant/polymer and polymer flooding from "Poor" to "Preferred." Another example is the affect of viscosity on *in situ* combustion (see Figure 9-5). A sharp change occurs, from "Poor" to "Good," as the viscosity increases. Another sharp change occurs, from "Good" to "Not Feasible," as the viscosity increases further. Even though these changes are relatively sharp, they are not as crisp as those shown in Figures 9-4 to 9-14 or as used as those in the crisp expert system.

The scenarios in Example 4 demonstrate yet another related problem. If we add information about salinity and porosity to Example 1 so that we can use all of Figures 9-4 to 9-14, and if we change the viscosity and the gravity and composition to be consistent with the heavier oil viscosity, we can demonstrate the *in situ* combustion and surfactant/polymer viscosity problems.

Example 4   - Scenario One
(1)   Gravity = 15 degrees API
(2)   Viscosity = 999 cp
(3)   Composition = high percent of C_5 - C_{12}
(4)   Salinity = 50,000 ppm
(5)   Oil saturation = 50%
(6)   Formation type = sandstone (homogeneous)
(7)   Payzone thickness = 35 ft
(8)   Average permeability = 1000 md
(9)   Well depth = 2000 ft
(10)   Temperature = 110°F
(11)   Porosity = 28%

Scenario Two
(1)   Gravity = 15 degrees API
(2)   Viscosity = 1001 cp
(3)   Composition = high percent of C_5- C_{12}
(4)   Salinity = 50,000 ppm
(5)   Oil saturation = 50%
(6)   Formation type = sandstone (homogeneous)
(7)   Payzone thickness = 35 ft

(8) Average permeability = 1 000 md
(9) Well depth = 2000 ft
(10) Temperature = 110°F
(11) Porosity = 28%

The difference between these two scenarios is only 2 centipoise or 0.2% in viscosity. If we list the rankings of the top four methods computed from Scenario One, we find *in situ* combustion ranked second and surfactant/polymer ranked fourth.

Scenario One (Rankings)
    1- Steam flooding   132 points
    2- *In situ* combustion   125 points
    3- Alkaline flooding   117 points
    4- Surfactant/polymer   116 points

Scenario Two (Rankings)
    1- Steam flooding   132 points
    * *In situ* combustion   65 points (Not Feasible)
    2- Alkaline flooding   117 points
    *- Surfactant/polymer   66 points (Not Feasible)

With only the small change in viscosity (2 centipoise), the *in situ* combustion and surfactant/polymer techniques drop from the second and fourth ranked methods to ones that are Not Feasible. Even though a rather sharp drop in feasibility occurs, the small viscosity increase makes a drop this sharp seem unreasonable.

Examples Three and Four demonstrate the kinds of problems experienced by some expert systems decision boundaries. Although there are several ways to reduce these problems, the problem of screening EOR methods is ideally suited to fuzzy logic. Fuzzy logic is like human logic at those boundaries. Instead of deciding which side to be on, a weighted average of each side is used. This makes the transition from one side of the boundary to the other much smoother. The fuzzy logic approach is discussed in the next section.

An important task of the expert system is to give, the user meaningful advice about the individual EOR methods on the basis of the raw scores computed by the program. For these expert systems, the raw scores were normalized on the basis of a maximum possible best score of 100% for the best possible process, which is steam flooding. That is, if all methods were to receive the best possible score, steam flooding would get the highest score, with 148 points. It also has the largest number of "Preferred" ratings of the methods shown in Figures 9-4 to 9-14. The other EOR methods (except the microbial drive) are all rated quite close to the steam flooding method. The raw score of 148 corresponds to 100%. All raw scores are divided by 148 to produce a normalized score relative to the best score possible.

At the end of a session, the scores are tallied, providing the user with a ranked list of candidates to take to the next screening step as well as an idea of how good the candidates are relative to the best possible score. So far in these examples, both expert systems have given realistic results, except in those cases where the fuzzy decision was

Chap. 9   Fuzzy Logic for Enhanced Oil Recovery   199

important. These expert systems have been run using much of the information given in Ref. 8 for actual EOR projects. In about 60% of the cases run, the method ranked highest by the expert system was the method that was actually selected and used in that project. In most of the other cases, the actual method used was ranked in the top three by the expert system. This is not too unusual because the actual test data influenced the scores used by the expert system. Expert systems are often built by comparing the results of the expert system with the results given by the experts, then modifying the system until it is as good as the experts. This approach gives us confidence in the accuracy of the results predicted by the expert systems.

9.3   EXPERT SYSTEMS AND FUZZY LOGIC

Most texts on artificial intelligence and expert systems [9,10] point out that almost all real expert systems have to deal with some kind of uncertainty. In building the first (crisp) expert system, considerable effort was expended examining the literature and working with raw data to reduce the uncertainty. If, for example, an EOR method gets a "Good" rating for an EOR criterion, that rating is assumed with 100% confidence, to be worth 10 points. Considerable effort went into defining the boundaries of the various ratings within each EOR criteria. For the crisp expert system, each rating block is considered to be a crisp set, that is, either the EOR method gets a particular rating or it doesn't. For instance (see Figure 9-4), if the API gravity of the oil is greater than 40, the hydrocarbon miscible method gets a "Preferred" rating. If the gravity is not greater than 40, the method gets some other rating. This works fine as long as the gravity is not near the boundary (in this case 40). But if it is, then some uncertainty arises. For example, what if the gravity is 27, right about the boundary between "Good" and "Poor" for the hydrocarbon miscible method? Should the score be 0 for "Poor" or 10 for "Good"? The crisp expert system makes a decision and assigns a membership, to either "Poor" or "Good" for the hydrocarbon miscible gravity, and the score for the hydrocarbon miscible method is incremented appropriately.

The fuzzy expert system reduces the uncertainty caused by set boundaries by replacing the crisp sets with fuzzy sets [11]. Fuzzy logic is conventional logic, or inference rules, that is applied to fuzzy sets rather than crisp sets. Fuzzy sets are represented by membership functions. Unlike the crisp sets, the value for an EOR criterion for an EOR method can have membership in more than one set. Figures 15-23 are the membership functions, or fuzzy sets, that correspond to the crisp sets in Figures 9-4 to 9-14.

There are no corresponding fuzzy sets for Figure 9-6 (oil composition), or Figure 9-9 (formation type). These two EOR criteria remained crisp for this study. Note that some of the abscissae of the fuzzy sets are different from those shown for the corresponding crisp sets. In these cases, simple transformations were used on the EOR criteria variable to better fit them to the fuzzy expert system shell.

As can be seen by observing Figures 9-15 to 9-23, each of the values for each of the EOR criteria, for each EOR method has a membership in more than one fuzzy set. If we continue our example with the API gravity and observe Figure 9-15a for the

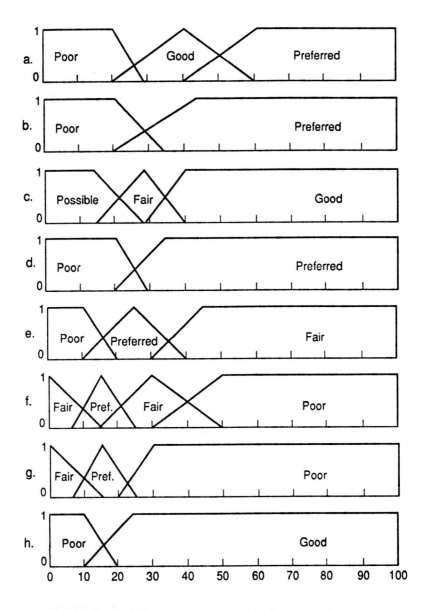

Figure 9-15. Membership functions for gravity for the:
(a) Hydrocarbon Miscible Method; (b) Nitrogen and Flue Gas Method;
(c) Carbon Dioxide Method; (d) Surfactant/Polymer and Polymer Flooding
Methods; (e) Alkaline Flooding Method; (f) *In Situ* Combustion
Method; (g) Steam Flooding Method; and (h) Microbial Drive
Method.

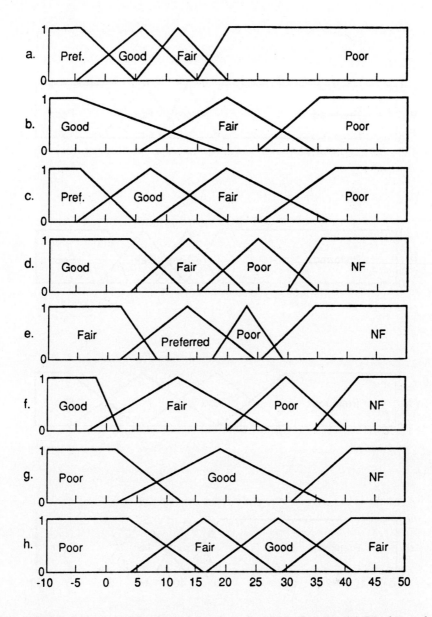

Figure 9-16. Membership functions for viscosity for the (a) Hydrocarbon Miscible Method, (b) Nitrogen and Flue Gas Method, (c) Carbon Dioxide Method, (d) Surfactant/Polymer and Polymer Flooding Methods, (e) Polymer Flooding Method, (f) Alkaline Flooding Method, (g) *In Situ* Combustion Method, and (h) Steam Flooding Method.
NF = Not Feasible

201

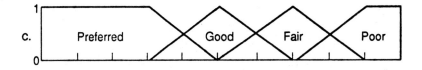

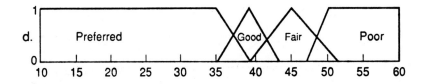

Figure 9-17. Membership functions for salinity for the
(a) Surfactant/Polymer Method, (b) Polymer Flooding Method,
(c) Alkaline Flooding Method, and (d) Microbial Drive Method.

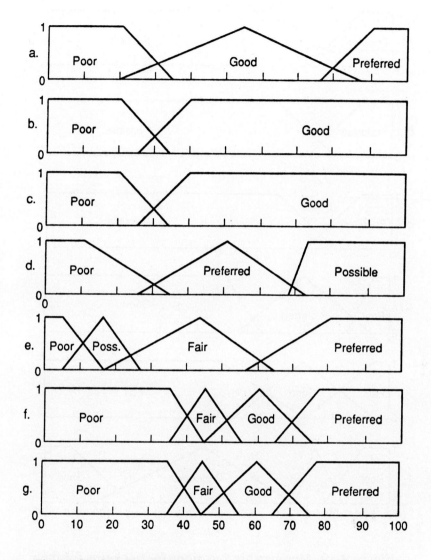

Figure 9-18. Membership functions for oil saturation for the (a) Hydrocarbon Miscible Method, (b) Nitrogen and Flue Gas Method, (c) Carbon Dioxide Method, (d) Surfactant/Polymer Method, (e) Polymer Flooding Method, (f) *In Situ* Combustion

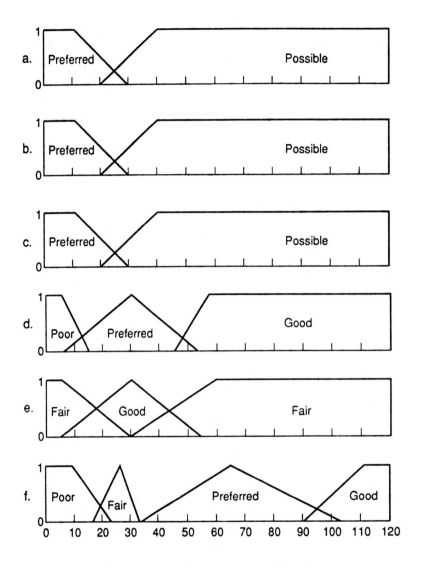

Figure 9-19. Membership functions for thickness for the (a) Hydrocarbon Miscible Method, (b) Nitrogen and Flue Gas Method, (c) Carbon Dioxide Method, (d) Surfactant/Polymer Method, (e) *In Situ* Combustion Method, and (f) Steam Flooding Method.

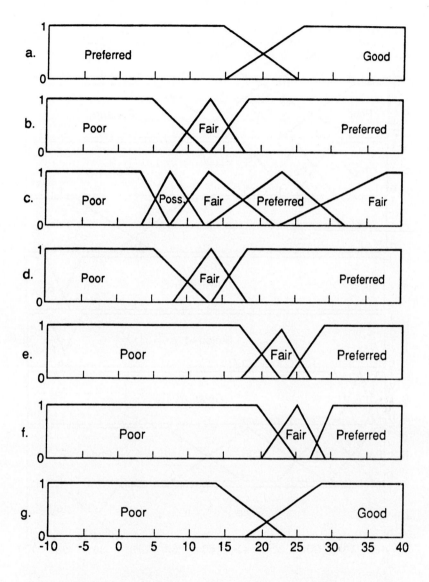

Figure 9-20. Membership functions for permeability for the (a) Hydrocarbon Miscible Method, (b) Surfactant/Polymer Method, (c) Polymer Flooding Method, (d) Alkaline Flooding Method, (e) *In Situ* Combustion Method, (f) Steam Flooding Method, and (g) Microbial Drive Method.

205

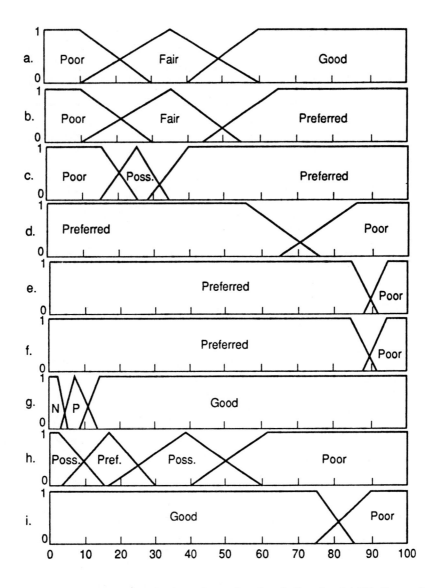

Figure 9-21. Membership functions for depth for the (a) Hydrocarbon Miscible Method, (b) Nitrogen and Flue Gas Method, (c) Carbon Dioxide Method, (d) Surfactant/Polymer Method, (e) Polymer Flooding Method, (f) Alkaline Flooding Method, (g) *In Situ* Combustion Method, (h) Steam Flooding Method, and (i) Microbial Drive Method.

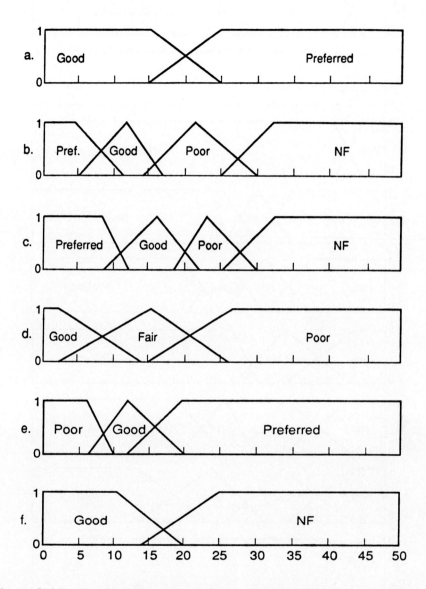

Figure 9-22. Membership functions for temperature for the (a) Nitrogen and Flue Gas Method, (b) Surfactant/Polymer Method, (c) Polymer Flooding Method, (d) Alkaline Flooding Method, (e) *In Situ* Combustion Method, and (f) Microbial Drive Method.
NF = Not Feasible.

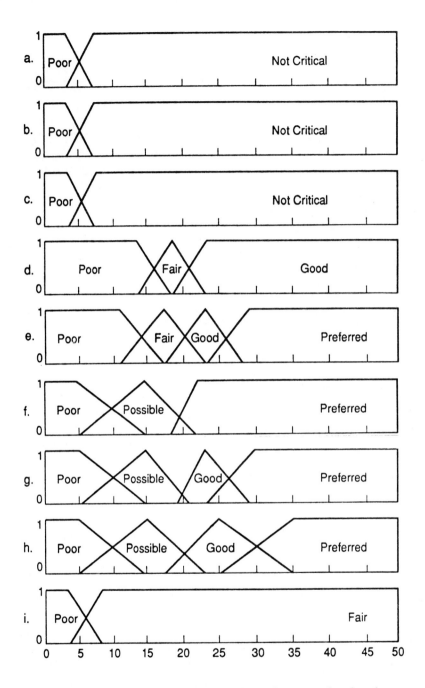

Figure 9-23. Membership functions for porosity for the
a) Hydrocarbon Miscible Method, (b) Nitrogen and Flue Gas Method,
(c) Carbon Dioxide Method, (d) Surfactant/Polymer Method, (e) Polymer
Flooding Method, (f) Alkaline Flooding Method, (g) *In Situ* Combustion
Method, (h) Steam Flooding Method, and (i) Microbial Drive Method.

Chap. 9   Fuzzy Logic for Enhanced Oil Recovery   209

hydrocarbon miscible method, we see that for a gravity of 27 the hydrocarbon miscible method has a membership of about 0.3 in "Poor" and a membership of about 0.3 in "Good". These memberships are combined to produce a crisp score. Our example demonstrates how memberships are combined to produce a crisp score.

Since a gravity of 27 for the hydrocarbon miscible method has membership in two sets, two rules are fired, each with a "strength" relative to the set membership value (in this case 0.3 for each rule). The two rules are:
1. If gravity_Hydrocarbon_Miscible is Poor
    Then Score = Poor
2. If gravity_Hydrocarbon_Miscible is Good
    Then Score = Good.

Figure 9-24 shows the membership functions for the output or the Score. From the rules above we can see that the Score should be part "Good" and part "Poor" resulting in a crisp value somewhere between 0 and 10. There are several methods for combining memberships. The one used by our fuzzy expert system is called the Max-Min Inference Method. This method combines the "Good" and "Poor" Scores by clipping the output membership function triangles at the height of the membership function value. (in this case the height is 0.3 for both "Good" and "Poor".) The crisp value for the Score is the centroid of the combination of these two truncated triangles. In this case it is the integer value 4. This procedure for returning a crisp value from a fuzzy calculation is known as the centroid defuzzification method. Figure 9-25 is a composite drawing of a portion of Figure 9-15a and a portion of Figure 9-24. It shows how the input and output membership functions are connected by the rules and how the crisp output Score is computed based on the number of rules fired and the value of the membership function for the rule premises. (e.g., in this case the membership function value for each premise for each rule was 0.3.)

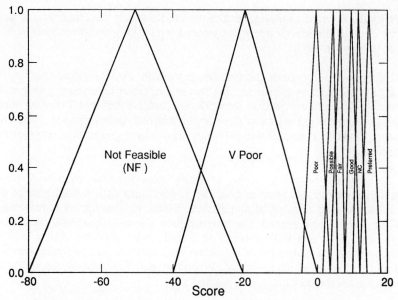

Figure 9-24. Output membership functions for the score.

210   Chap. 9   Fuzzy Logic for Enhanced Oil Recovery

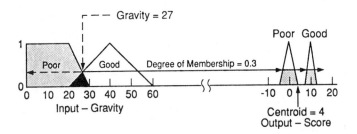

Figure 9-25. Demonstration of the max-min inference method.

9.4   HOW THE EXPERT SYSTEMS WORK

If an engineer were solving the EOR screening problem by hand, using the backward-chaining or goal-driven method, he would first pick a goal (for example, the hydrocarbon gas injection method from the left-hand side of Tables I and II). The engineer would then pick the subgoals that would have to be met before the original goal could be satisfied (for example, the gas injection category.) This process of picking subgoals would continue as long as necessary, but in our case, it would stop here. The engineer would ask only those questions necessary to determine whether gas injection would be a feasible category. If the feasibility of the gas injection category were established, the engineer would ask only those questions necessary to determine whether the hydrocarbon method would be feasible. If not, another goal would be picked. If yes, the problem would be solved, unless more than one solution were desired, in which case, another goal would be picked and the process continued.

With the forward-chaining, or data-driven, approach, the engineer lets the data help search through the search tree (the system keeps asking questions until it is clear which node to move to next).

The crisp expert system, the first one assembled, uses backward-chaining. With this system, the approach is to first assume that hydrocarbon injection is going to work. In order for hydrocarbon injection to work, the category of gas injection must be applicable. In order for gas injection to be applicable both the oil property data and the reservoir data shown in Figures 9-4 to 9-14 must have scores greater than preprogrammed threshold values.

The program begins by trying to verify these subgoals by asking questions about gravity, viscosity, oil composition, etc. It continues until a final goal is met or until an assumption is rejected at some level. When an assumption is rejected, that branch of the search tree is pruned. The program then moves to the next unpruned branch to the right and picks that EOR process as a goal, then continues until a solution is found. Since we want a ranked list of candidate EOR methods, the program searches the tree until all possible solutions are found. When the search is finished, the solutions are printed, with a score for each qualifying method.

Chap. 9   Fuzzy Logic for Enhanced Oil Recovery   **211**

Figure 9-26 is a portion of an *and/or graph* for a portion of the search space for the crisp version of the expert assistant. It is called an *and/or graph* because the branches connected by an arc are *and* branches (all of the leaves must be true, and in this case, must have a preprogrammed minimal score, before the branch is resolved). The unarced branches are *or* branches. They require only a single truth (minimal score) for resolution.

The fuzzy expert system was written next. It uses forward-chaining and, essentially, an exhaustive search. It starts with the API gravity of the oil in the reservoir (Figure 9-15) and assigns a score to each EOR method. It then moves on to viscosity (Figure 9-16) and repeats the procedure. The procedure is repeated until all 11 EOR categories are checked. The fuzzy expert system actually uses some crisp rules, combined with the fuzzy rules. Figure 9-6 shows oil composition screening data. Figure 9-9 shows the screening data for the reservoir rock formation type. In both of these cases there was not enough information to fuzzify the data. So for this study the rules for these two figures remained crisp.

Another area where the rules remain crisp is one in which an EOR criterion offers no options for an EOR method. Figure 9-7 shows screening data for formation salinity. For five of the nine EOR methods formation salinity is not critical. This gives rise to five crisp rules.

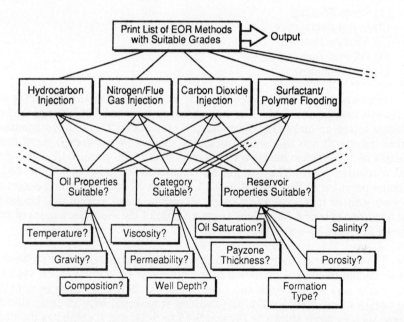

Figure 9-26. *And/or* graph for a portion of the search space for the CLIPS backward-chaining version of the problem.

One difference between the two expert systems is the tools, or expert system shells, used. Each system uses a different shell, as is discussed in the next section.

9.5   PROGRAM COMPARISONS AND SUMMARY

The crisp expert system was written with the expert system shell, CLIPS [12], developed by NASA. CLIPS is a forward-chaining shell written in the C programming language. It is a very versatile and flexible shell, which can even be used to write expert systems in the backward-chaining mode, as was done for the crisp expert system (backward-chaining was used because it is more intuitive and, therefore, easier to prune search trees).

The crisp expert system is a great improvement over the hand calculation method that utilizes graphs and charts. Considerable information has been added to the expert system, as can be seen in Figures 9-4 to 9-14. A final example of this is the first example problem in this paper in which two conditions have been added from Table II. The salinity is 50,000 ppm and the porosity is 28%. Using this information with Table II one would get the same solution we obtained in our sample session, as shown in Figure 9-2 and described in the text. This example, again, shows that the only method that can be used is steam flooding. The expert assistant, however, produces a ranked list of five different candidate processes. They are, in order, as follows:

	Score (%)
(1) Steam flooding	89
(2) *In situ* combustion	85
(3) Alkaline flooding	76
(4) Polymer flooding	73
(5) Microbial drive	72

The expert system has provided the solutions to the two problems we had earlier, when using only Table I. It has given us a ranked list, instead of just one candidate or a large unranked list of candidates. Methods such as *in situ* combustion can be ranked because it can also weigh problems such as "What does it mean to have a temperature of 110°F when the table says greater than 150°F preferred"? and it gives the method a relative score. This weighting is possible because of all the additional information provided in Figures 9-4 to 9-14. As pointed out earlier, this expert system works very well on most real world cases. Examples 3 and 4 point out, however, that there is a definite potential for serious errors because of the sharp boundaries of the crisp sets shown in Figures 9-4 to 9-14.

The fuzzy expert system was written to eliminate this potential problem and to add some human-like fuzzy reasoning to the otherwise rigid crisp expert system. This expert system was written with the Togai Fuzzy C development system (13). This system does a lot of work for the programmer; it makes it easy to enter membership functions, such as those shown in Figures 9-15 to 9-23, and it computes the necessary centroids, as demonstrated in Figure 9-25. This system shell is harder to use than CLIPS because the programmer must write a C language program to drive the Fuzzy C program. This means that the programmer has to write the search routines and other peripheral management software that is typically already supplied with shells like CLIPS. Although this allows more flexibility, a great deal of time is required to write search routines with the sophistication of those found in CLIPS. Because it was easiest to write, a

Chap. 9 Fuzzy Logic for Enhanced Oil Recovery   213

forward-chaining exhaustive search was used on this expert system. Still, extensive coding was required.

This expert system does a much better job on problems such as those discussed in Examples 3 and 4. In Example 3, the crisp expert system causes dramatic changes between the two Scenarios, even though the input data for the two scenarios are very similar. The results shown in the ranked list presented with Example 3 are from the crisp expert system. The following results are from the fuzzy expert system.

Scenario One (Rankings)
1-	Alkaline flooding	109 points
2-	Polymer flooding	107 points
3-	Surfactant/polymer	101 points
4-	Carbon dioxide	97 points
5-	Microbial drive	89 points
6-	Hydrocarbon miscible	86 points
7-	*In situ* combustion	83 points
8-	Nitrogen and flue gas	82 points
9-	Steam flooding	81 points

Scenario Two (Rankings)
1-	Alkaline flooding	112 points
2- (tie)	Polymer flooding	109 points
3- (tie)	Surfactant/polymer	109 points
4-	Carbon dioxide	102 points
5- (tie)	Microbial drive	89 points
6- (tie)	Hydrocarbon miscible	89 points
7- (tie)	*In situ* combustion	87 points
8- (tie)	Nitrogen and flue gas	87 points
9-	Steam flooding	78 points

Only small changes occur between Scenarios One and Two when the fuzzy expert system is used. In fact the only changes are small changes in the total points awarded. The relative rankings are not really changed.

Example 3 is intended to be a realistic problem, but it is a worst case. The overall raw scores or points produced in the fuzzy version of Example 3 show little increase from Scenario One to Scenario Two. This means that the predicted viability of the EOR methods will not be unduly enhanced by small changes in the input data by the fuzzy expert system.

In Example 4 (Scenario One) the crisp expert system ranked *in situ* combustion as the second best method and surfactant/polymer as fourth best. In Scenario Two, the only change in the input data was an increase of 0.2% in the oil viscosity, hardly a measurable change. This change caused the *in situ* combustion and surfactant/polymer methods to be discarded. They were "Not Feasible." The fuzzy expert system keeps *in situ* combustion as the second best method and surfactant/polymer as the fourth best method in both scenarios, partly because the abscissae shown in Figure 9-5 and used in the crisp expert system were converted to a logarithmic scale and plotted linearly in Figure 9-16.

This is how they are used in the fuzzy expert system. The transformation equation is as follows: transformed-viscosity = (integer) (10*$\log_{10}$ (viscosity) + .5). (The scale shown in Figure 9-6 is linear data plotted on a logarithmic graph.) The transformation itself tends to fuzzify the set boundaries. The transformation was made because the fuzzy expert system shell doesn't handle very large numbers or long scales very well. The fuzzy membership functions are meant fuzzify the set boundaries. But when any other output membership function is combined with the "Not Feasible" output membership function, with its centroid at -50, it's hard to make the result of the boundary change very gradual. The -50 score was designed to dramatically reduce the raw score of an EOR method that was thought to be "Not Feasible". This is a good idea if the criterion value is not near the set boundary. Even though a change in feasibility may be quite dramatic as the criterion value changes, it most likely is not a step function. Complete resolution of this problem will require a little more work.

The fuzzy expert system is much better at solving problems, such as those in Examples 3 and 4, than the crisp expert system is. Although these "worst case" problems do not represent the majority of EOR screening problems, they are real, and some degree of the crisp set boundary problem is present in almost every EOR screening problem. Our crisp expert system works more like a classical expert system than the fuzzy expert system does. The crisp system works interactively with the user. It tries to prune the search tree and it offers a simple explanation facility. On the other hand, with the fuzzy expert system, users enter the data and wait for all of the scores to be computed. If the users want some explanation, they can request a dump and watch the progress of the score calculation.

Some of the differences between the two expert systems occur because fuzzy expert systems are designed to fire all the rules that apply to the problem, even those that have only a minor influence on the outcome. A conventional expert system, like the crisp expert assistant does just the opposite, that is, it tries to prune the search tree by eliminating any consideration of rules that have little or no influence on the problem outcome.

The final issue we will discuss is the development of the membership functions for the fuzzy sets shown in Figures 9-15 to 9-24. Reference 13 states that, "Determining the number, range, and shape of membership functions to be used for a particular variable is somewhat of a black art." It further states that trapezoids and triangles, such as those shown in Figures 9-15 to 9-24, are a good starting point for membership functions. Trapezoids and triangles served as a starting point for membership functions for this project. The membership functions in Figures 9-15 to 9-24 are still trapezoids and triangles but many of them are different from those used as the starting points. Some effort was spent polishing the membership functions and several changes were made. In many cases the changes made little difference in the final scores, but in some cases they made a great deal of difference. Ideally, we would expect the triangular membership functions to resemble bell-shaped curves and the trapezoids to resemble S-shaped curves. References 10 and 14-16 suggest methods for determining better membership functions. Example 4 shows that, in at least some cases, there is a need for improved membership functions. Improving the membership functions will require taking a harder look at the available data and will be the subject of another study. The idea of using neural nets, fuzzy pattern recognition, or genetic algorithms [16] to "teach" the membership functions to improve their shape is intriguing and should be considered for a future project.

REFERENCES

[1]   Stosur, J. G. ,"The Potential of Enhanced Oil Recovery," *International Journal of Energy Research, Vol.* 10, 357-370 (1986).

[2]   Taber J. J. and F. D. Martin, "Technical Screening Guides for Enhanced Recovery of Oil," paper presented at the 58th Annual Society of Petroleum Engineers Technical Conference, San Francisco, California, October 5-8, 1983 (SPE 12069)

[3]   Goodlet, G. O., H. M. Honarpour, H. B. Carroll, and P. S. Sarathi, "Lab Evaluation Requires Appropriate Techniques-Screening for EOR-I," *Oil and Gas Journal,* 47-54 (June 23, 1986).

[4]   Goodlet, G. O., H. M. Honarpour, H. B. Carroll, P. Sarathi, T. H. Chung, and D. K. Olsen, "Screening and Laboratory Flow Studies for Evaluating EOR Methods," Topical Report DE87001203, Bartlesville Project Office, USDOE, Bartlesville, Oklahoma, November 1986.

[5]   Parkinson, W. J., G. F. Luger, R. E. Bretz, and J. J. Osowski, "An Expert System for Screening Enhanced Oil Recovery Methods," paper presented at the 1990 Summer National Meeting of the American Institute of Chemical Engineers, San Diego, California, August 19-22, 1990.

[6]   Donaldson, E. C., G. V. Chilingarian, and T. F. Yen, (Editors), *Enhanced Oil Recovery, I Fundamentals and Analysis* (Elsevier, New York, 1985).

[7]   Poettmann, F. H., (Editor), *Improved Oil Recovery* (The Interstate Oil Compact Commission, Oklahoma City, Oklahoma, 1983).

[8]   "Enhanced Recovery Methods are Worldwide" (Petroleum Publishing Company, 1976). (Compiled from issues of *The Oil and Gas Journal*).

[9]   Luger, G. F. and W. A. Stubblefield, *Artificial Intelligence and the Design of Expert Systems* (The Benjamen/ Cummings Publishing Company, Inc., Redwood City, California, 1989).

[10]   Giarratano, J. C. and G. Riley, *Expert Systems - Principles and Programming* (PWS - Kent Publishing Company, Boston Massachusetts, 1989).

[11]   Parkinson, W. J., K. H. Duerre, J. J. Osowski, G. F. Luger, and R. E. Bretz, "Screening Enhanced Oil Recovery Methods with Fuzzy Logic," paper presented at the Third International Reservoir Characterization Technical Conference, Tulsa Oklahoma, November 3-5, 1991.

[12]   Giarratano, J. C., *CLIPS User's Guide, Version 4.3*, Artificial Intelligence Section, Lyndon B. Johnson Space Center, June 1989.

[13]   Hill, G., E. Horstkotte, and J. Teichrow, *Fuzzy-C Development System User's Manual, - Release 2.1*, Togai Infralogic, Inc., June 1989.

[14]   Turksen, I. B., "Measurement of Membership Functions and Their Acquisition," *Fuzzy Sets and Systems* (40) 538 (Elsevier Science Publishers B. V., North-Holland, 1991).

[15]   Klir, G. J. and T. A. Folger, *Fuzzy Sets, Uncertainty, and Information* (Prentice Hall, Englewood Cliffs, New Jersey, 1988).

[16]   Karr, C., "Genetic Algorithms for Fuzzy Controllers," AI Expert, February 1991, pp. 2633.

10

A FUZZY LOGIC RULE-BASED SYSTEM FOR PERSONNEL DETECTION

Paul Sayka
Los Alamos National Laboratory

An approach which is based on fuzzy logic for a personnel detection system in a special facility is proposed. The system is composed of an array of various electrical sensors which are used to detect if a human being is present in a particular area. In this chapter, a fuzzy rule-based system is compared with a crisp rule-based system for a hypothetical model system. With the implementation of fuzzy logic, a system that is difficult to implement in traditional techniques becomes more feasible.

10.1　INTRODUCTION

Detection schemes to identify whether or not a person is in a specific location are currently available. These schemes include so-called "electric eyes," infrared detectors, sonar transmitters, a host of other transducers, and even human operators monitoring a remote area with cameras. None of these schemes is particularly efficient, most are prone to error, and even the system utilizing operators relies on an alert, attentive operator.

An efficient detection system would open doors to a vast expanse of new products. Trains and buses would only make stops at locations where passengers are detected. Security systems could monitor areas to verify that there are no intruders present. Climate control in a room or office would depend on whether the space was occupied. Hotels could install detectors in rooms, and in the event of a fire, the hotel management would know which rooms contained guests at the time of the fire, allowing a speedy and organized rescue. One can envision many disaster situations, such as an earthquake, where the knowledge of a persons location is crucial information. Clearly, the development of a personnel detection system would benefit mankind.

The personnel detection system can be implemented with a fuzzy rule-based system that allows the designer to specify less accurate sensors, which in turn are less expensive. The fuzzy rule-based system is more reliable, more adaptive, and provides more accurate information as compared with the crisp or traditional rule-based system that utilizes precise values.

10.2 FUZZY LOGIC PRIMER

Fuzzy logic is a mathematical tool that allows modern digital computers to model systems that are imprecisely defined. This technology allows the computer to make a decision from vague or imprecise data since it does not require the precise specifications required by traditional computing techniques.

A fuzzy logic based system reduces the inaccuracy that occurs at crisp or traditional set boundaries by substituting fuzzy sets. A set consists of a collection of elements. Crisp sets have membership requirements that restrict the value of the set element to either a zero or a one. The boundaries of the crisp set are clearly defined. One corresponds to complete membership in the crisp set and zero corresponds to null membership in the set. On the other hand, in a fuzzy set, a set element can be a member of the set with a partial degree of membership between zero and one. There is a gradual transition from membership to nonmembership in a fuzzy set and not the sharply defined boundaries as in the crisp set. The grade of membership in the fuzzy set is expressed by a membership function, the most commonly used function being a triangular shape. Linguistic variables or "labels" may be used to represent a fuzzy set. For example, variables such as "hot," "cold," and "warm" could represent different fuzzy sets for temperature. Modifiers such as "very," "almost" or "mostly" may be added to form fuzzy subsets such as "very hot" or "mostly warm."

One of the most popular implementations for fuzzy logic is in the area of control systems. Production rules for control applications are used to map input variables to output variables using IF-THEN statements. The conditions or antecedents of the rule follow the word IF and the conclusion or consequence follows the word THEN. The rule has the general form,

IF (condition), THEN (action)
input   ===>   output

The ith. control rule appears as follows:

IF X1 is Ai, X2 is Bi, X3 is Ci,. . ., THEN Y is Oi

where Xn are the inputs, Y is the output, Ai, Bi, and Oi are membership functions, and n is the number of antecedents. The rules governing the system produce outputs depending upon which rules are in effect and the measure of belief in the premise of each rule. Multiple input systems may exist and the system may have several rules "firing" in parallel at the same instant to produce a single output. The firing strength of the ith. rule is used to determine a control decision.

Membership functions are combined using an inference method based on the compositional rule of inference for approximate reasoning suggested by Zadeh (1973). The inference is used to derive a conclusion from implications (facts) and premises. The final fuzzy output level is found using the Max-Min or Max-Dot inference techniques, which are the most popular methods of inference.

The number of fuzzy inferences per second is denoted as "fips." From a signal processing point of view, high speed waveform generators are used to produce time displaced triangular waveform patterns for the respective membership functions. The important characteristics of the pulse such as the height and width of the triangular pulse as well as the time displacement of the pattern are controlled by the pattern generator. The generator runs at typical clock frequencies in the megahertz range for modern digital computers.

The fuzzy output that results from inference must be "defuzzified" to obtain a crisp numerical result for the physical output device. The most popular method of defuzzifying is the centroid or center of gravity method. In this method, the centroid for the clipped (max-min) or scaled (max-dot) area of the fuzzy inference is calculated to produce a fixed value that can be crisp output to the physical device. The block diagram of the complete process from input to output with two rules being fired is shown in Figure 10-1. The action is executed with the degree of membership of the rules condition. These rules are modified by fuzzy operations such as AND, OR, and NOT. These operations allow the antecedents to be combined into premises. Given two antecedents a and b, the fuzzy operations are defined as follows:

a AND b => min.(a,b), minimum of a and b

a OR b => max.(a,b), maximum of a and b

NOT a,b => 1.0 - a,b

For example, the expression (Temperature is High) OR (Pressure is Low) would be evaluated as the maximum of {(Temperature is High), (Pressure is Low)} where the membership function for High and Low gives the degree of membership of Temperature and Pressure, respectfully, in each set.

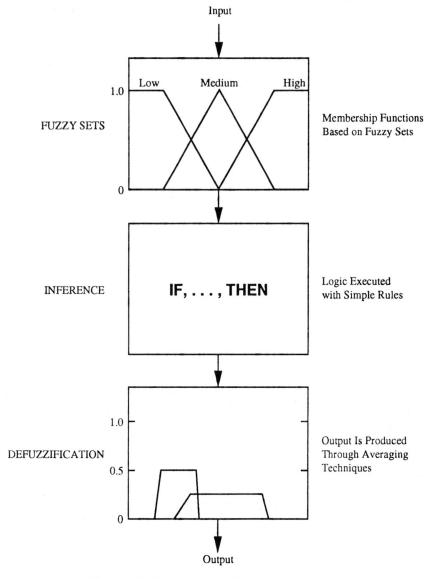

Figure 10-1. Fuzzy logic control process.

In summary, a fuzzy decision is the result of weighing the evidence and its importance in the same manner that humans make decisions. Fuzzy logic reflects humanlike thinking where the human can deduce an imprecise conclusion from a collection of imprecise premises.

Chap. 10 Fuzzy Logic for Personnel Detection

10.3 SENSORS

A sensor is an electrical transducer, a device capable of translating input energy of one form into an output energy of another form. In most cases, the output of the transducer is a electrical signal which is then used by some other device which provides information or control functions. Sensors provide precise or crisp data. A wide variety of transducers are available and the selection depends upon the application. Some possible sensors that could be employed to detect a human presence include:

- Motion Sensors
- Audio Sensors
- Metal Detectors
- Optical Sensors
- Infrared Detectors

There are a great deal more sensors that one could consider, but in general the more sophisticated the detection principle, the more expensive the device.

The sensor arrangement to be examined in this chapter involves two different types of the above sensors. The application involves detecting a human presence as he/she enters and exits a room. The arrangement of the sensors is shown in Figure 10-2. The integrated system includes two sets of optical sensors located near the door of the room and an infrared (IR) detector. These sensors are further detailed in the next two sections.

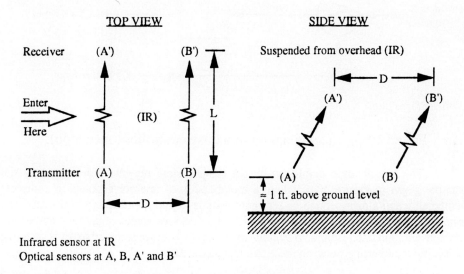

Figure 10-2. System configuration.

Infrared Detector

An inexpensive infrared detector has already been employed in an air conditioner that is controlled with fuzzy logic. The air conditioner, manufactured by Mitsubishi Heavy Industries Ltd., utilizes an infrared sensor to recognize if people are in the room being cooled. The controller then adjusts the power level of the unit according to whether anyone is in the room, realizing a savings on power bills by an average 24%. A similar sensor may be employed for the personnel detection scheme developed in this chapter.

The infrared detector used in a crisp system would simply sense a temperature and deliver that information to a controller which compares the measured value to some setpoint and acts accordingly if an error signal exists. The error signal being defined as the difference between the setpoint and the measured parameter. A crisp system such as this relies on control rules like the following,

IF Eo (error) > 0, THEN CHANGE OUTPUT PROPORTIONALLY

IF Eo = 0, THEN DO NOTHING

A block diagram of an open-loop crisp controller is shown in Figure 10-3. To form a closed-loop feedback controller, the input of the sensor is tied to the output of the crisp controller.

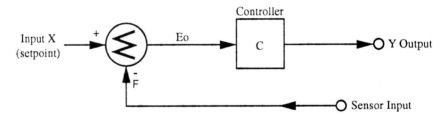

Error Signal, Eo = X - F

Figure 10-3. Block diagram of a crisp controller (open loop).

The binary logic employed by the crisp system may be represented by the example given in Figure 10-4, where a membership of one corresponds to complete membership and a membership of zero corresponds to null membership in the set. The system has only two states and the system is not defined for values of partial membership between zero and one. Below 96 degrees and above 104 degrees the crisp set is zero, the set has full membership between these two temperatures.

Chap. 10 Fuzzy Logic for Personnel Detection 223

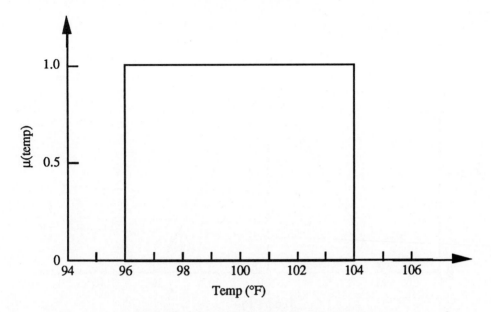

Figure 10-4. Crisp set representation.

The same sensor used for the crisp system can be used in a fuzzy rule-based system, but the control rules are modified. For example, let the setpoint of the IR detector system be 96 degrees Fahrenheit. In the crisp system, a person in the vicinity of the sensor is recognized only if the sensor sees 96 degrees or more. The system will act if it sees 96 degrees or more, but will do nothing if it sees 95 degrees. The crisp system uses the familiar two-valued, true-false logic known as binary logic. Most real world devices generate crisp data. Fuzzy data consists of an array of believability values, each between zero and one, this allows any degree of truth to be represented, rather than just the binary 1 or 0. In the fuzzy system, membership functions are applied to different values of the measured parameter. For example, a fuzzy set labeled "hot" for the IR detector might look as follows,

$$\frac{0.00}{96} + \frac{0.25}{97} + \frac{0.50}{98} + \frac{0.75}{99} + \frac{1.00}{100} + \frac{0.75}{101} + \frac{0.50}{102} + \frac{0.25}{103} + \frac{0.00}{104}$$

The numbers in the upper positions are the degrees of membership that correspond to fuzzy set temperatures that are in the lower position. The "+" symbol denotes a union operation. Note that this set is "normal" since at least one member of the set has a membership value of one. The membership value for 98 degrees reflects a measure of belief of 50 percent. This fuzzy set may also be represented graphically as shown in Figure 10-5.

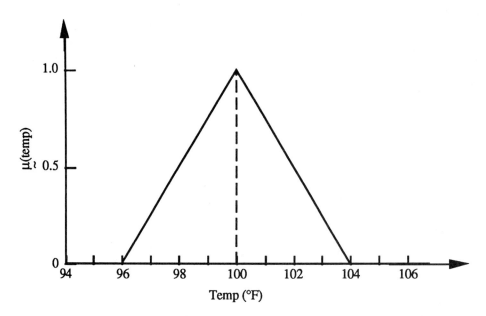

Figure 10-5. Fuzzy set representation.

Optical Sensors

An inexpensive optical sensor has already been employed in a washing machine that is controlled with fuzzy logic. The washing machine, manufactured by Matsushita, utilizes two optical sensors that determine how dirty the clothes are and the size of the laundry load. The machines fuzzy controller then chooses an appropriate cleaning cycle out of the multitude that are available from memory. The same company employs optical sensors in a new fuzzy vacuum cleaner. In this machine, the volume of dust present is measured and the controller acts to regulate the vacuum cleaner motor speed. A similar inexpensive sensor may be employed for the personnel detection scheme developed in this paper.

The detector used in a crisp system would sense the absence of light and send the information to a crisp controller which compares the abnormal situation with normal situation (setpoint). An error signal is generated when there is a difference between the setpoint and the current situation.

The arrangement of the optical sensors in this system are configured to provide both speed and direction information (if so desired) on the person who has entered the sensor array area. The direction information is obtained by noting which set of optical sensors was triggered first. This crisp information can be represented in a truth table as follows:

Chap. 10 Fuzzy Logic for Personnel Detection

SENSOR A	**SENSOR B**	**LOGIC OUTPUT**
0	0	NO ACTION
0	1	EXITING AREA
1	0	ENTERING AREA
1	1	ONE IN, ONE OUT (or a system fault)

where A and B are transducers and zero indicates no person detected and one corresponds to a human being detected.

The optical sensors can act as a velocity (motion) sensor. Motion is detected at one sensor and the transit time (delta T) between sensors being activated is derived and used to compute the speed from the distance (D) and the time elapsed. A crisp system would operate such that if speed greater than some crisp setpoint is detected, then there is a human present. In the crisp case, if the speed is slightly below the setpoint the system fails to detect a human presence. On the other hand, a fuzzy system could use a fuzzy set to describe the conditions. An example of a fuzzy set named "fast" that applies in this case might be the following:

$$\frac{0.00}{5 \text{ mph}} + \frac{0.50}{6 \text{ mph}} + \frac{1.00}{7 \text{ mph}} + \frac{1.00}{8 \text{ mph}} + \frac{1.00}{9 \text{ mph}}$$

The average speed of a human at a normal walking pace is about 4 miles per hour (mph). This fuzzy set may be represented graphically as a slope ramping up to a membership value of one from left to right. At seven mph, the membership function is a flat shoulder that continues to the right.

There are other sensor configurations that may be used for the application specified in this chapter. For instance, the optical sensors could have been utilized in a manner that measures the amount of time there is no light present between the transmitter and receiver sensors. That is, between A and A' or B and B' as shown in Figure 10-2. This gives information on the width of the item passing before the sensors if a constant speed is assumed. Figure 10-6 will clarify this application.

The time the light is not sensed at the receiver (delta T prime) is proportional to the thickness (W) of the object. A new fuzzy set may be formed where the maximum membership function is assigned to the average width of a persons leg at a point one foot above the ground level. Objects which enter the sensor field will be compared with data to determine whether or not the object has width characteristics that are similar "or close to" the thickness of a average persons lower calf area of the leg. A fuzzy set labeled "normal" that conveys this information has the following form,

$$\frac{0.00}{6} + \frac{0.5}{5} + \frac{1.00}{4} + \frac{0.5}{3} + \frac{0.00}{2}$$

226   Chap. 10   Fuzzy Logic for Personnel Detection

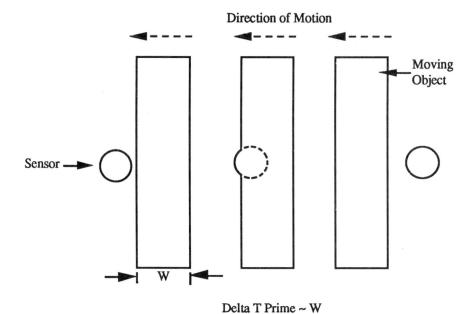

Figure 10-6. Alternate sensor arrangement.

where the number below the line is the average width of a persons leg in inches at a point one foot above ground level.

10.4   SENSOR ACCURACY

The output of an actual transducer is affected by the nonideal behavior of the transducer, which causes the indicated measurand to deviate from the true value. The measurand is the quantity, or condition that is being measured. The error is the algebraic difference between the indicated value and the true value of the measurand. Accuracy is defined as the ratio of the error to full-scale output. Crisp data is subject to error.

Fuzzy rules can also reduce the uncertainty surrounding a failure of the input transducer(s). The same fuzzy sets such as "low" and "high" can be used in rules to deduce conclusions regarding the operational status of one or both of the sensors. Assume two identical sensors labeled A and B are used to obtain two inputs for the proposed system. The effect of a single sensor failure is governed by one of the following rules:

If sensor A is low and sensor B is normal, Then use the higher sensor,
If sensor A is normal and sensor B is high, Then use the higher sensor,
If sensor B is normal and sensor B is too high, Then use the lower sensor.

If both of the transducers fail, one of the following rules is in effect:

If sensor A is low and sensor B is low, Then error,
If sensor A is low and sensor B is high, Then error,
If sensor A is high and Sensor B is high, Then error.

10.5 CONTROL RULES

Fuzzy control rules or production rules in the rule-base are of the form:

If X is A1 and Y is B1, then Z is C11;
If X is A1 and Y is B2, then Z is C12;

.
.
.

If X is Ai and Y is Bj, then Z is Cij;

The statement that follows IF is the condition (or premise) and the statement following THEN is the action (or conclusion). The input variables are X and Y, Z is the output variable.

These typical control rules may be applied to the topic of this chapter in the following manner. Let X be the input from the infrared sensor and Y be the input from the combined optical sensors. Ai, Bi, Cij can be represented by linguistic variables, they are fuzzy subsets of X, Y, and Z. Each Ai, Bi, or Cij has a universe of discourse, for example, a fuzzy set "med" or "medium" for the IR detector might have values in the range [98, 102].

Other fuzzy subsets for the IR detector input might be represented as shown in Figure 10-7. In this case we interpret medium as "a temperature near about 100 degrees Fahrenheit. Note that overlap of about 25 percent is suitable for ensuring multiple rules fire. As the amount of overlap increases, more rules will fire. Little or no overlap causes erratic or jerky control. Fuzzy subsets of the speed input from the optical sensor array might look like Figure 10-8.

Control rules that map the input space to the output space are:

IF temp is very low,	AND speed is slow,	THEN Z is very low
IF temp is very low,	AND speed is medium,	THEN Z is low
IF temp is very low,	AND speed is fast,	THEN Z is very low
IF temp is low,	AND speed is slow,	THEN Z is low
IF temp is low,	AND speed is medium,	THEN Z is medium
IF temp is low,	AND speed is fast,	THEN Z is low

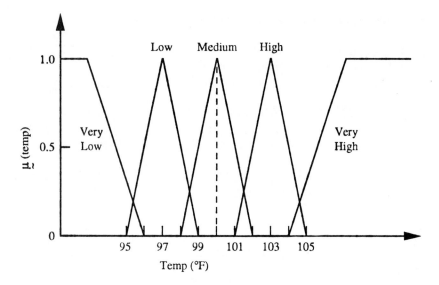

Figure 10-7. IR detector fuzzy sets.

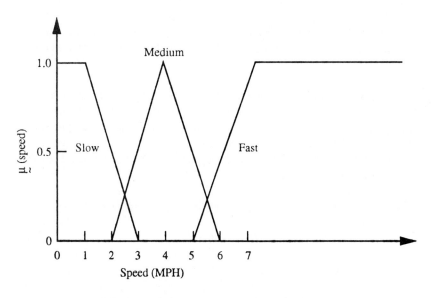

Figure 10-8. Optical sensor fuzzy sets.

Control rules that map the input space to the output space are:

IF temp is very low,	AND speed is slow,	THEN Z is very low
IF temp is very low,	AND speed is medium,	THEN Z is low
IF temp is very low,	AND speed is fast,	THEN Z is very low
IF temp is low,	AND speed is slow,	THEN Z is low
IF temp is low,	AND speed is medium,	THEN Z is medium
IF temp is low,	AND speed is fast,	THEN Z is low
IF temp is medium,	AND speed is slow,	THEN Z is medium
IF temp is medium,	AND speed is medium,	THEN Z is high
IF temp is medium,	AND speed is fast,	THEN Z is medium
IF temp is high,	AND speed is slow,	THEN Z is low
IF temp is high,	AND speed is medium,	THEN Z is medium
IF temp is high,	AND speed is fast,	THEN Z is low
IF temp is very high,	AND speed is slow,	THEN Z is very low
IF temp is very high,	AND speed is medium,	THEN Z is low
IF temp is very high,	AND speed is fast,	THEN Z is very low

The output variable "Z" is the control variable that is used to initiate a graded response or action that is based upon the number and strength of the rules that fired. The output fuzzy set requires "defuzzification" to produce a crisp electrical control signal that will be used by physical devices as a measure of the belief that there is a human in the sensor area.

10.6 CONFLICT RESOLUTION

A conflict is defined as a situation where the state of the system cannot be determined. In the case of the crisp system, this might occur when two sensors are providing contradictory information. This is possible in the crisp system because the sensors operate independently from one another and conflicting solutions can exist. For example, for the system proposed in this chapter, the optical sensors may have detected a person in the room, but the IR sensor has not detected a heat energy level that is high enough to cause the controller to act. In order to remedy this situation several schemes may be employed.

A coincidence scheme may be used to resolve the problem of conflicting information from sensors in the crisp case. The basic idea is to use an odd number of sensory inputs, so that a tie situation can be resolved. A simple example might employ three sensors and operate such that before the controller will act on an error signal > 0, two of the three (2/3 logic) sensors must be in agreement. This scheme has the obvious disadvantage of requiring more sensors.

Another scheme includes adding extra control rules to cover the situations where a conflict has occurred. Thus, to compensate for these conflict situations, the crisp rule-based system will have more rules and require more computer time and memory as compared with the fuzzy rule-based system. The crisp system would require many rules such as

"If temp is 100 degrees and sensor A is < 80 degrees, Then error"

or

"If temp is 100 degrees and sensor A is > 120 degrees, Then error"

Contradictory data can be accepted by the fuzzy system since the ambiguity of the crisp set boundary has been eliminated. For the fuzzy rule-based case, overlapping of membership functions ensures that more than one rule applies to each value of the input. This in turn makes the control of the system smoother than in the crisp case. Overlapping rules also mean that there is a fuzzy rule in effect under all conditions and in most cases, several rules are fired at a given time so that conflict situations will not occur.

10.7   CONCLUSION

The fuzzy rule-based personnel detection system has several distinct advantages over a similarly configured crisp rule-based system. Fuzzy controllers work in a way a human might when controlling a system by taking into account several factors simultaneously and applying common sense. Only unambiguous phenomena can be investigated properly through the application of crisp logic principles.

In the fuzzy system, conflict situations are easily resolved, whereas in the crisp system, conflict can cause the system to hang-up and not provide an output. The fact that a variable is a member of contradictory sets is unacceptable in most conventional forms of logic, but is an important feature of fuzzy logic. Fuzzy logic can handle processes that must weigh multiple inputs or have conflicting constraints.

In the crisp system, the sensors and system must be tuned or calibrated for optimum performance. Setpoint drift or measurement error can severely affect the crisp system. In the fuzzy system, sensor measurement error and mistuning of the system are not as critical as in the crisp system. Less accurate sensors may be utilized with the fuzzy rule-based system as evidenced in their increased use in moderately priced home appliances that are flooding the Japanese market as they exploit fuzzy logic. Less expensive sensors result in a cost savings and the savings could be used to procure more sensors which would increase reliability. That is, extra sensors may be used as back-up or redundant sensors which would maintain the system integrity even if a sensor were to fail.

The fuzzy system requires less control rules, and this means less computer memory is required, and also the speed of the system is faster than the crisp counterpart since the control rules are fired in parallel in the fuzzy system. Reduced data handling in fuzzy systems means less software and hardware. Typically the rulebase of the fuzzy system is an order of magnitude smaller than a similar knowledge-based system. Also the fuzzy rules involve less math and are in simpler in nature because they use everyday language.

From a signal processing standpoint, reduced buss switching means less rise and fall transitions for the electronic components which reduces propagation delay through the circuitry and hence increases speed of processing. The fuzzy system is more adaptive than its crisp counterpart. In the fuzzy case, automatic controller adjustment in real time is possible to continually optimize the performance of the control system.

The next generation of fuzzy personnel detectors might include neural networks which could train the system on an evolving basis, providing a more robust system. By Using neural nets the personnel detector could learn to be more efficient and become more personalized by tuning its membership functions under changing conditions. Hardware and software tools are now available to configure fuzzy logic rule-based systems.

REFERENCES

Business Week, May 21, 1990 (pg. 92-93).

Computer Design, March 1, 1991 (pg. 90-94, 97-100, 102).

Discover, January, 1991 (pg. 60-61).

Klir, G. J., and T. A. Folger. *Fuzzy Sets, Uncertainty, and Information*. Prentice Hall, Englewood Cliffs, New Jersey, 1988

Norton, H. N. *Handbook of Transducers*. Prentice Hall, Englewood Cliffs, New Jersey, 1989.

Togai Infralogic, Inc. *Fuzzy-C Development System User's Manual*, Release 2.1., 1989.

Zadeh, L. A. "Outline of a new approach to the analysis of complex systems and decision processes." IEEE Transactions on Systems, Man, and Cybernetics, 3, 28-44, 1973.

11

USING FUZZY LOGIC TO AUTOMATICALLY CONFIGURE A DIGITAL FILTER

Robert J. Knight
Mohammad T. Akbarzadeh
University of New Mexico

A fuzzy logic based self-tuning filter is simulated and implemented using the DSP 56001 digital signal processor IC and an IBM PC. The fuzzy logic tuner uses a hierarchy of fuzzy rules to decide on the upper and lower edges of the filter. These edges are then used to design a Butterworth band pass filter. The coefficients of the transfer function of the Butterworth filter are passed on to the DSP 56001 chip for implementation of the filter.

11.1 INTRODUCTION

A common problem in signal processing is the extraction of a desired signal from extraneous noise or interference. This problem is relatively easy to handle when the spectral characteristics of the signal and noise are known beforehand. However, it becomes more difficult when either or both of these parameters is unknown.

Methods exist for automatically configuring filters to extract the signal from the noise, such as adaptive filtering, in which the filter modifies its own parameters so as to minimize some error function.

This paper introduces a new fuzzy logic approach to the design of a self-tuning filter. The self-tuning bandpass filter uses a fuzzy logic rule-based algorithm to determine the upper and lower bounds of the frequency band of the signal of interest. These bounds are then used to compute the coefficients of the transfer function of a bandpass Butterworth filter. The fuzzy algorithm takes the fast Fourier transform of the frequency band and computes the power spectrum of the received signal. By observing the slope and level of the power spectrum at each frequency, it creates two fuzzy membership functions which are used to determine the upper and lower edges of the filter. The value of these confidence functions at each frequency describes the likelihood that that frequency is an edge.

Section 2 details the theory behind the rule-based algorithm. A set of fuzzy rules with membership functions are presented in this section. Section 3 contains simulations of the filter. Varying types of frequency bands is applied to the self-tuning filter, and the filter is shown to find approximate edges of the frequency band of the signal of interest. A discussion about performance and feasibility of the fuzzy-logic rule-based algorithm in signal processing applications follows in Section 5.

11.2 THEORY OF OPERATION

When an expert looks at a frequency spectrum of a given noisy signal and needs to determine which band of frequencies are 'good' and which are noise, he/she refers to his/her past experience with signals and decides on where the 'good' signal is located in the frequency spectrum. The approach in fuzzy logic is to formulate what goes through that expert's mind into a set of rules. These rules in combination with the membership functions of its variables construct our knowledge-based fuzzy-logic expert analyzer. If one can draw a complete model of the expert's brain activity using a sequence of fuzzy interrelated rules, then hypothetically, one will be able to replace the expert with its electronic counterpart.

To construct this fuzzy-logic knowledge based model, one needs to have access to an expert to assist in creating the knowledge-base using his experience. Table 11-1 shows the set of fuzzy rules used to determine the upper and lower edges of the signal of interest. The central theme of the rules may be expressed as follows:

> Edges usually happen where *Level* is not too high and not
> too low and the *Slope* is not too much negative and
> not too little negative,

where *Level is* the normalized level of the frequency spectrum at frequency fi, and *Slope* is defined to be the slope of the spectrum coming from the left of the frequency index to

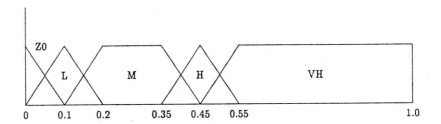

Figure 11-1. Level Membership Functions.

right. The above rule is expanded to cover all possible combinations of low and high levels, and slopes of the spectrum. Each frequency is given a confidence number which is the membership value of that frequency to the fuzzy concept of an "edge."

Table 11-1: Lower Band-Edge Determination Rules

		Level					
		Z0	VL	L	M	H	VH
	NBIG	VL	VL	VL	VL	VL	VL
	NMED	VL	VL	VL	L	L	VL
Slope	Z0	VL	VL	VL	L	VL	VL
	PMED	VL	VL	M	VH	L	VL
	PBIG	VL	VL	L	VL	VL	VL

Membership functions for fuzzy variables such as Zero, Low, Medium, High, and Very High *Level*, Negative Big, Negative Medium, Zero, Positive Medium and Positive Big *Slope*, and Very Low, Low, Medium, High, and Very High *Confidence* need to be defined. Please see Figures 11-1 to 11-3 for drawings of the membership functions. Given these membership functions, the fuzzy rules in Table 11-1 were constructed. These rules apply to the low band-edge detector; the high band-edge detector rules are identical except the sign of the slope membership functions are reversed in the rulebase.

To determine where the lower edge is, every frequency from the lower bound of the power spectrum to the peak value of the frequency spectrum is analyzed. Each individual fuzzy rule will produce a level of confidence. The output of all rules are then inferred using the centroid method to determine the membership of the frequency fi in the fuzzy concept of *Confidence*. The centroid of this fuzzy concept is then taken to determine the lower edge of the filter.

For the upper edge of the filter, a similar procedure is followed with the exception that the range of analysis is from the peak frequency to the upper end of the frequency spectrum.

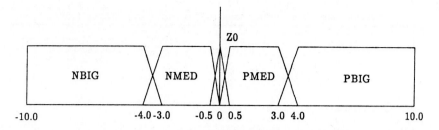

Figure 11-2. Slope Membership Functions.

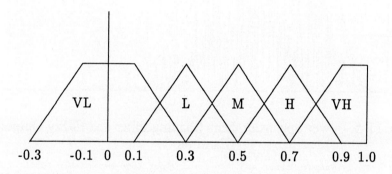

Figure 11-3. Confidence Membership Functions.

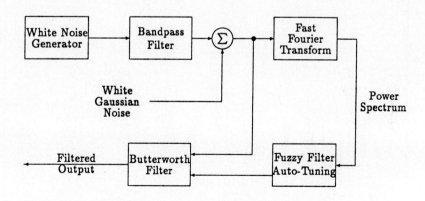

Figure 11-4. System Block Diagram.

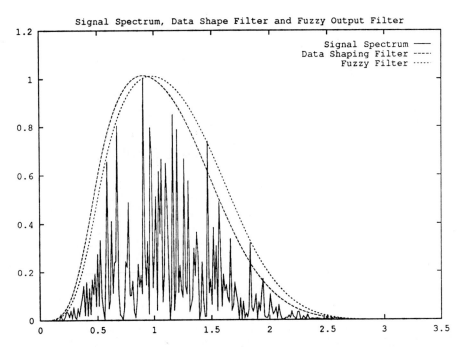

Figure 11-5. Power Spectrum, Data Shaping Filter and Fuzzy Output Filter (Example 1).

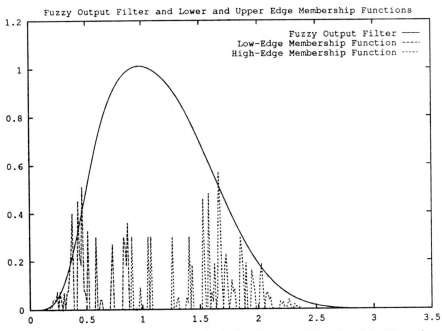

Figure 11-6. Fuzzy Output Filter and Confidence Membership Functions (Example 1).

Chap. 11   Fuzzy Logic to Configure a Digital Filter   **237**

11.3   SIMULATION

The simulation of the fuzzy-logic bandpass self-tuning Butterworth filter has three steps. The first step is to generate an input signal which contains a band-limited signal of interest corrupted by white Gaussian noise. The second step is to feed this noisy signal into the fuzzy band-edge estimation algorithm. The third step uses the fuzzily estimated edge frequencies to generate a bandpass Butterworth filter suitable for removing the noise. A block diagram of the simulation software is shown in Figure 11-4. Listings of some of the programs used are included in the appendices.

The first step produces the signal corrupted by noise using software for the simulation. A random number generator was used to create white Gaussian noise. Then, the white noise was fed through a bandpass Butterworth filter to generate the signal of interest. A white noise of smaller magnitude was then added to the signal of interest.

The simulation then takes the power spectrum of the corrupted signal. In order to determine the lower edge of the frequency band of interest, it then steps through all frequencies below the maximum-power frequency, creating a function describing the confidence that a given frequency is the lower edge of the frequency band of interest. The confidence function is defuzzified to determine the estimated location of the lower edge.

The simulation then repeats the process to determine the upper edge of the frequency band.

After finding the edges, the simulation computes the coefficients of a bandpass Butterworth filter that will pass the signal of interest while filtering out the interfering noise.

Figures 11-5 and 11-6 illustrate this process. Figure 11-5 shows the power spectrum for a data process filtered between digital frequencies $\omega L = 0.5$ and $\omega H = 1.5$. (*Note: All frequencies in this report are expressed in terms of digital frequency, which extends from 0 [corresponding to DC] to π, [corresponding to half the sampling frequency (sometimes called the Nyquist frequency).]*) No external noise was added to the data signal. Shown along with the power spectrum is the frequency profile of the filter used to shape the signal of interest, as well as the frequency profile of the filter designed by the fuzzy algorithm. The filter profiles can be seen to be quite close in this case.

Figure 11-6 shows the frequency profile of the filter designed by the fuzzy algorithm together with the *Confidence* membership functions for the upper and lower band edges.

Figures 11-7 and 11-8 illustrate the same data process, but with a signal-to-noise ratio of 5. In Figure 11-7, the fuzzy algorithm together with the *Confidence* membership functions determine the fuzzy-designed filter's profile which can be seen to be slightly broader than the original data-shaping filter, but still within acceptable limits. Figure 11-8 shows the fuzzy output filter profile together with the *Confidence* membership functions and the input's filter profile. The increased noise causes the *Confidence* functions to be non-zero for a broader range of frequencies, forcing the output filter to be wider.

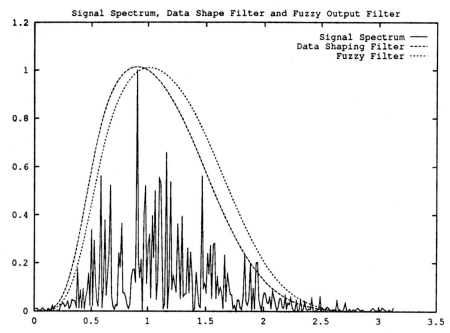

Figure 11-7. Power Spectrum, Data Shaping Filter and Fuzzy Output Filter (Example 2).

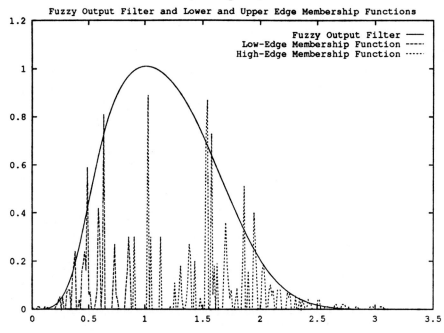

Figure 11-8. Fuzzy Output Filter and Confidence Membership Functions (Example 2).

Chap. 11 Fuzzy Logic to Configure a Digital Filter

The third example illustrates a narrowband data process (extending from $\omega L = 0.1$ to $\omega H = 0.3$), with a signal-to-noise ratio of 10. Figure 11-9 illustrates the power spectrum as well as the input and output filter profiles.

Figure 11-10 illustrates the original data signal, before the corrupting noise was added, together with the same signal after the noise was added. Figure 11-11 shows the original data signal with the signal available at the output of the fuzzy-designed filter. The fuzzy-filtered signal can be seen to be significantly closer to the original than the unfiltered signal.

11.4 HARDWARE IMPLEMENTATION

11.4.1 Details of Implementation

The implementation of the fuzzy-logic bandpass self-tuning Butterworth filter has five steps. The first step is to collect the sampled signal data from the DSP 56001. This is done via an RS-232 link between the DSP 56001 and an IBM PC. The next step is to generate the frequency spectrum of the incoming signal using the fast Fourier transform. Once the frequency spectrum is generated, the next step is to apply the fuzzy algorithm to estimate the upper and lower edges of the required filter. The fourth step is to compute the coefficients of the Butterworth bandpass filter using these fuzzily estimated edge frequencies, and the last step is to send those coefficients to the DSP 56001 and filter the incoming data.

Figure 11-12 depicts the operation of the system in block diagram form. Lists of the programs used by the IBM-PC are included in the Appendices.

The first step in the fuzzy edge-determination algorithm is to acquire and sample the input signal. This is done by the DSP 56001 chip, a digital signal processor with a 24-bit internal data bus. The data are acquired using 16-bit analog-to-digital converters controlled by the DSP 56001. The sampling rate is 13945 Hz, so the frequency analysis will be valid for frequencies between DC and 6972 Hz. Once the required number of data samples have been collected, they are transmitted to the IBM PC via a 2400 baud RS-232 link.

The next step produces the frequency spectrum of the incoming signal. Once it has all the acquired data, the IBM PC computes the power spectrum of the signal using the fast Fourier transform. The fuzzy edge-determination algorithm then begins. In order to determine the lower edge of the frequency band of interest, it steps through all frequencies below the maximum-power frequency, creating a function describing the confidence that a given frequency is the lower edge of the frequency band of interest. The confidence function is defuzzified to determine the estimated location of the lower edge.

The fuzzy algorithm then repeats this process, stepping upward from the frequency with the maximum power to determine the upper edge of the frequency band.

240 Chap. 11 Fuzzy Logic to Configure a Digital Filter

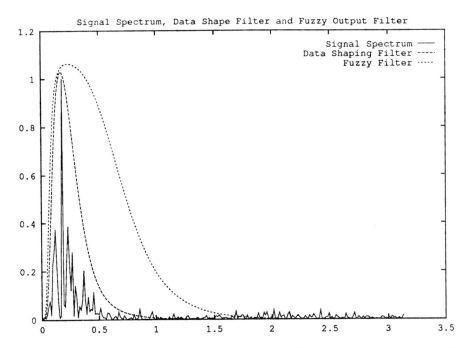

Figure 11-9. Power Spectrum, Data Shaping Filter and Fuzzy Output Filter (Example 3).

After finding the edges, the fuzzy algorithm computes the coefficients of a first-order bandpass Butterworth filter that will pass the signal of interest while filtering out the interfering noise.

The last step is to download the filter coefficients to the DSP 56001 chip and use them to implement the Butterworth filter.

11.4.2 Experimental Results

The DSP 56001 fuzzy filter-autotuning system was experimentally tested on a variety of input signals, including a sinusoid, a speech signal and music.

The spectrum of a 2000 Hz sinusoid is shown in Figure 11-13, together with the fuzzy filter derived from it. When presented with data samples describing the sinusoid, the fuzzy filter autotuning system created the confidence functions shown in Figure 11-14. These confidence functions defuzzify to band-edges of 1890.3 and 2126.6 Hz; these band-edge frequencies were used to design the filter shown in Figures 11-13 and 11-14.

Figure 11-15 illustrates an approximately 20 msec sample of speech. Figure 11-16 shows the power spectrum of the speech signal and the frequency profile of the filter derived from it by the fuzzy-logic system. Figure 11-17 illustrates the fuzzy filter and the

Chap. 11 Fuzzy Logic to Configure a Digital Filter

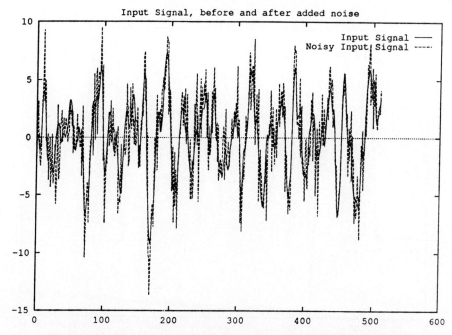

Figure 11-10. Clean and Noisy Input Signals (Example 3).

confidence functions created while designing it. For this input, the autotuning system estimated the band edges at 157.5 Hz and 1128.9 Hz.

The final example used a segment of music of approximately 20 msec duration as its input data. The power spectrum of the music is shown in Figure 11-18. The confidence functions fuzzily extracted from the power spectrum are depicted in Figure 11-19. These confidence functions defuzzify to band-edge frequencies of 183.8 and 1207.7 Hz. These frequencies were used to design the fuzzy filter illustrated in Figures 11-18 and 11-19.

Although the fuzzily-designed filters shown in the above figures are rather wide, this is because the fuzzy autotuning system presently uses only first-order bandpass filters. Higher-order filters would increase the sharpness of the filter profiles. As can be seen, the fuzzy autotuning technique is fairly successful at detecting the edges of frequency bands in various signal environments.

11.5   DISCUSSION

Although fuzzy logic is well suited to fields such as control systems, where a fuzzy rulebase will have a small number of unlike inputs (such as pressure, velocity, temperature, etc.), signal processing deals with a large number of variables, each representing a quantity very similar in nature to the others (for example, the 257 different variables representing the power spectrum in these simulations). In order to perform the

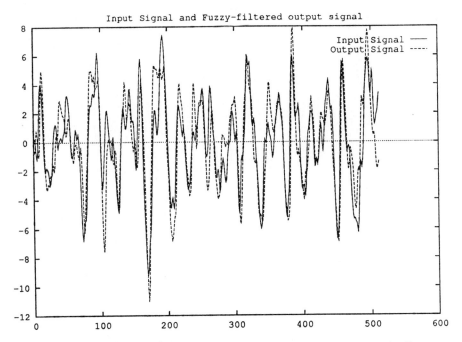

Figure 11-11. Clean and Fuzzy-Filtered Signals (Example 3).

band-edge detection entirely in fuzzy logic, a fuzzy-logic routine would need 257 fuzzy variables, each of whose membership functions must be specified separately, even if they are identical.

A common technique used in traditional signal processing for dealing with large numbers of like variables is iteration. This is the technique used in this simulation: instead of calculating the confidence membership function based on the entire spectrum at once, it calculates the confidence at each point before moving on to the next.

Iterative methods are not possible in pure fuzzy logic, however. This forces iteration to be performed by routines external to the fuzzy logic rulebase. Thus, for fuzzy logic to be successfully applied to signal processing problems, some external routines must be developed to interface between the fuzzy rulebase and the iterative world of signal processing.

11.6 CONCLUSION

A fuzzy logic rule-based self-tuning band-pass Butterworth filter was designed, simulated, and tested. The results were satisfactory: in most cases the fuzzy algorithm was able to closely approximate the upper and lower edges of the filter. This was done using only 15

Chap. 11 Fuzzy Logic to Configure a Digital Filter

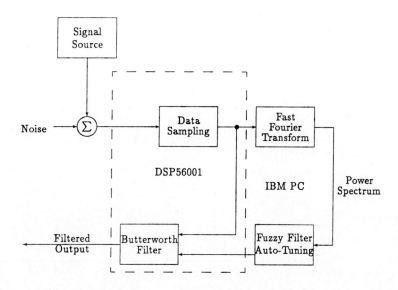

Figure 11-12. System Block Diagram.

rules which was a rather crude approximation of the human cognition. However, the performance can be greatly improved if one can express the full expertise and knowledge of an expert on filter theory in a set of rules. Fuzzy rules (as compared with Boolean crisp logic rules) enable the designer to better approximate human cognition, and therefore allows the designer to model the expert's expertise more accurately.

The successful application of fuzzy logic theory to signal processing and filter theory shows that research in the area of a combination of signal processing and fuzzy logic is promising. However, it can also be a difficult task because signal processing is a well developed, mathematically rigorous field that does not 'fuzzify' easily.

Despite the unpredictable, broadband nature of audio signals as examined in this report, the fuzzy autotuning system performed well at estimating the frequency band of interest. However, to adequately filter signal from noise on a continuous basis, the fuzzy autotuning system would need to run on a more or less continuous basis, updating the filter coefficients continually, instead of as a single initial estimation. For more predictable situations, however, such as detecting broadcasting stations, the fuzzy autotuning filter would perform well after only a single 'look' at the spectral environment.

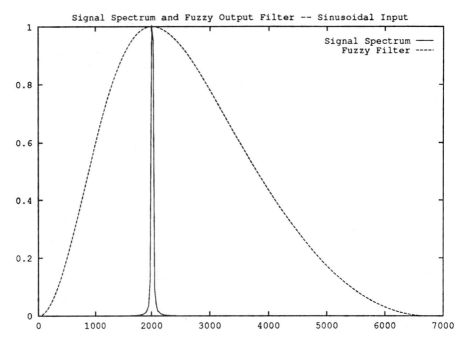

Figure 11-13. Sinusoidal Input Spectrum and Fuzzy Filter.

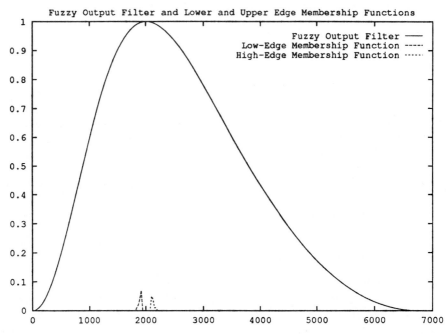

Figure 11-14. Fuzzy Filter and Confidence Functions for Sinusoidal Input.

Chap. 11   Fuzzy Logic to Configure a Digital Filter   **245**

APPENDIX A
HANDLING ROUTINES FOR FUZZY-LOGIC FUNCTIONS

```
#define FSAMP 13495.0

#include <math . h>
#include <stdio . h>
int round(float val);

float find_lower_edge(float *spectrum, int spec_length, int max_power_bin)
/* Use fuzzy estimation of how good a lower edge a given point is to estimate the lower
    edge of a frequency band. Input variables are array containing the power spectrum
    and the length of the spectral array. Returns the digital frequency of the lower edge
    of the filter (0 = DC; pi = Fsamp/2). */
{
        int current_bin;
        float local_avg, slope;
        int freq_index, edge_bin;
        signed char scaled_slope, *scaled_conf;
        unsigned char scaled_level;
        float *confidence;
        float sum_x, sum_x2, sum_xy;
        int points;

        FILE *conf_out;

        int window_width;   /* Width of averaging window at each step */

        conf_out = fopen("conf_lo", "wt");

        confidence = (float *) malloc(spec_length * sizeof(float));

        window_width = spec_length/100;

        if (max_power_bin == 0) {
/* Maximum power is at DC; output confidence function should be a singleton at DC */
confidence[0] = 1.0;
fprintf(conf_out, "0.0 1.0\n");
        } else {
        for (current_bin = 0; current_bin <= max_power_bin; current_bin++) {
                sum_x = sum_x2 = sum_xy = local_avg = 0.0;
                points = 0;
                for (freq_index = current_bin - window_width;
        freq_index <= current_bin + window_width;

        freq_index++) {
```

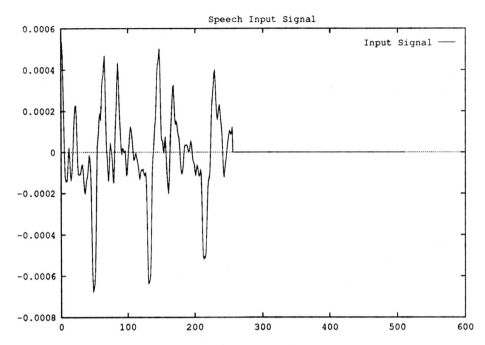

Figure 11-15. Speech Signal.

```
if ((freq_index >= 0) && (freq_index <= spec_length)) {
    local_avg += spectrum[freq_index];
    sum_x += (float) freq_index;
    sum_x2 += ((float) freq_index) * ((float) freq_index);
    sum_xy += freq_index * spectrum[freq_index];
    points++;
}
        }
        slope = ((points*sum_xy - sum_x*local_avg) * spec_length)
                / (points*sum_x2 - sum_x*sum_x);

        /* Clip slope to +/- 10 */
        if (slope > 10.0) {
            slope = 10.0;
        }
        if (slope < -10.0) {
            slope = -10.0;
        }
        local_avg /= 2.0 * window_width + 1.0;
        scaled_level = (unsigned char) 200 * local_avg;
        scaled_slope = (signed char) 10 * slope;
        Filt_Lo(scaled_level, scaled_slope, scaled_conf);
        confidence[current_bin] = (float) (*scaled_conf) / 100.0;
```

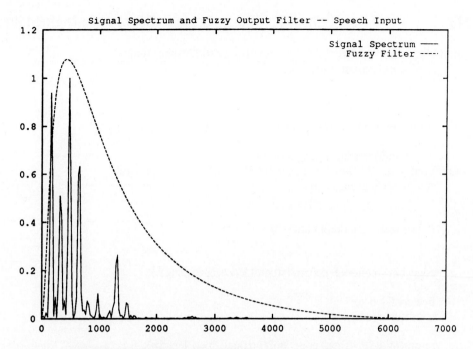

Figure 11-16. Speech Input Spectrum and Fuzzy Filter.

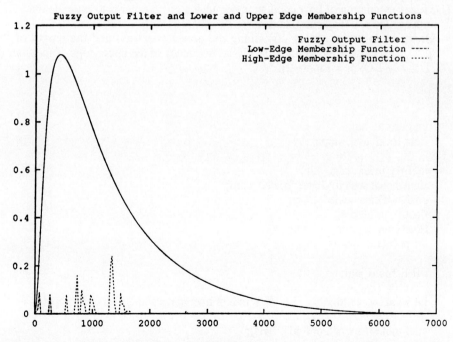

Figure 11-17. Fuzzy Filter and Confidence Functions for Speech Input.

248   Chap. 11   Fuzzy Logic to Configure a Digital Filter

```c
        fprintf(conf_out, "%f %f\n", (current_bin*FSAMP)/(2*spec_length),
            confidence[current_bin]);
        }
    }

        for (current_bin = max_power_bin+1; current_bin < spec_length; current_bin++) {
            confidence[current_bin] = 0.0;
    fprintf(conf_out, "%f %f\n", (current_bin*FSAMP)/(2*spec_length),
        confidence[current_bin]);
        }

        /* Defuzzify the output array */


        edge_bin = centroid_defuzzify(confidence, spec_length);

        free(confidence);
        fclose(conf_out);

        return(M_PI * ((float) edge_bin) / ((float) spec_length));
}
float find_upper_edge(float *spectrum, int spec_length, int max_power_bin)
/* Use fuzzy estimation of how good an upper edge a given point is to estimate the
    upper edge of a frequency band.
    Input variables are array containing the power spectrum and the length of the
        spectral array. Returns the digital frequency of the upper edge of the filter (0
        = DC; pi = Fsamp/2). */

{

        int current_bin;
        float local_avg, slope;

        int freq_index, edge_bin;
        signed char scaled_slope, *scaled_conf;
        unsigned char scaled_level;
        float *confidence;
        float sum_x, sum_x2, sum_xy;
        int points;

        FILE *conf_out;

        int window_width;          /* Width of averaging window at each step */

        conf_out = fopen("conf_hi", "wt");

        confidence = (float *) malloc(spec_length * sizeof(float));
```

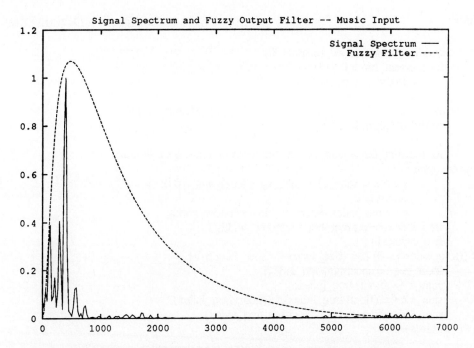

Figure 11-18. Music Input Spectrum and Fuzzy Filter.

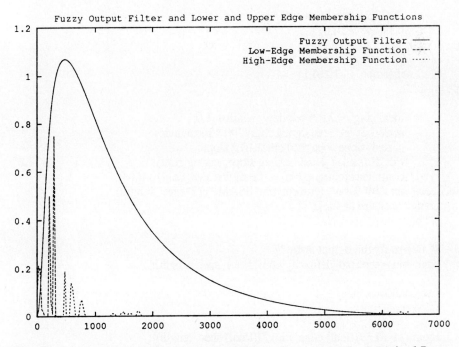

Figure 11-19. Fuzzy Filter and Confidence Functions for Musical Input.

```c
        window_width = spec_length/100;
        for (current_bin = 0; current_bin < max_power_bin; current_bin++) {
            confidence[current_bin] = 0.0;

fprintf(conf_out, "%f %f\n", (current_bin*FSAMP)/(2*spec_length),
    confidence[current_bin]);
        }
        for (current_bin = max_power_bin; current_bin < spec_length;
current_bin++) {
            sum_x = sum_x2 = sum_xy = local_avg = 0.0;
            points = 0;
            for (freq_index = current_bin - window_width;
    freq_index <= current_bin + window_width;
    freq_index++) {
if ((freq_index >= 0) && (freq_index <= spec_length)) {
    local_avg += spectrum[freq_index];
    sum_x += (float) freq_index;
    sum_x2 += ((float) freq_index) * ((float) freq_index);
    sum_xy += freq_index * spectrum[freq_index];
    points++;
}
            }
            slope = ((points*sum_xy - sum_x*local_avg) * spec_length)
                / (points*sum_x2 - sum_x*sum_x);

            /* Clip slope to +/- 10 */
            if (slope > 10.0) {
slope = 10.0;
            }
            if (slope < -10.0) {
slope = -10.0;
            }
            local_avg /= 2.0 * window_width + 1.0;
            scaled_level = (unsigned char) 200 * local_avg;
            scaled_slope = (signed char) 10 * slope;
            Filt_Hi(scaled_level, scaled_slope, scaled_conf);
            confidence[current_bin] = (float) (*scaled_conf) / 100.0;
fprintf(conf_out, "%f %f\n", (current_bin*FSAMP)/(2*spec_length),
    confidence[current_bin]);
        }

        /* Defuzzify the output array */
        edge_bin = centroid_defuzzify(confidence, spec_length);

        free(confidence);
        fclose(conf_out);

        return(M_PI * ((float) edge_bin) / ((float) spec_length));
}
```

Chap. 11 Fuzzy Logic to Configure a Digital Filter 251

```
int centroid_defuzzify(float mem_fctn[], int width)
/* Defuzzify an output fuzzy membership function (of size width, passed in the array
     mem_fctn) using the centroid method. Returns the index number of the centroid. */

{
        float sum_conf, sum_bin_conf;
        int index;

        sum_conf = sum_bin_conf = 0.0;
        for (index = 0; index < width; index++) {
                sum_conf += mem_fctn[index];
                sum_bin_conf += index * mem_fctn[index];
        }
        if (sum_conf == 0.0) {
return(width/2);
                } else {
return(round(sum_bin_conf / sum_conf));
                }
}

int round(float val)
/* Rounds a floating point number to the nearest integer. */

{
        if (fabs(val – (int) val) < 0.5) {
                return((int) val);
        } else {
                if (val < 0.0) {
return((int) (val – 1));
                } else {
return((int) (val + 1));
                }
        }
}
```

APPENDIX B
FUZZY-LOGIC RULEBASE FOR
LOWER-EDGE DETECTION

PROJECT Filt_Lo

/* SLOPE OF THE FILTER AT A GIVEN POINT.
* THE UNIVERSE OF DISCOURSE IS −10 TO 10.
* SLOPE IS MEASURED IN UNITS OF 0.1. */

VAR SLOPE
    TYPE SIGNED BYTE
    MIN −100
    MAX 100

    /* membership fuctions for Slope (NBIG, NMED, ZO, PMED, PBIG). */

    MEMBER ZO
        POINTS −5 0 0 1 5 0
    END

    MEMBER PMED
        POINTS 0 0 5 1 30 1 40 0
    END

    MEMBER NMED
        POINTS −40 0 −30 1 −5 1 0 0
    END

    MEMBER PBIG
        POINTS 30 0 40 1 100 1
    END

    MEMBER NBIG
        POINTS −100 1 −40 1 −30 0
    END

END /* END OF SLOPE DEFINITION. */

/* POWER LEVEL OF THE INPUT AT A GIVEN FREQUENCY,
    NORMALIZED WITH RESPECT TO THE MAXIMUM POWER.
    UNIVERSE OF DISCOURSE IS 0 TO 1; EXPRESSED IN TERMS OF
    0.005 NORMALIZED POWER UNITS. */

VAR LEVEL
    TYPE UNSIGNED BYTE
    MIN 0
    MAX 200
    /* Membership functions for LEVEL: ZO, VS, S, M, L, VL */

Chap. 11   Fuzzy Logic to Configure a Digital Filter   253

```
            MEMBER ZO
POINTS 0 1 20 0
            END

            MEMBER L
POINTS 0 0 20 1 40 0
            END

            MEMBER M
POINTS 20 0 40 1 70 1 90 0
            END

            MEMBER H
POINTS 70 0 90 1 110 0
            END

            MEMBER VH
POINTS 90 0  110 1  200 1
            END

      END /* End of LEVEL definition */
```

/* Confidence that a given point is an edge of the frequency band. The confidence vs. frequency curve forms the membership function for a fuzzy variable indicating the edge of the frequency band. Defuzzifying (by an external, non-Togai routine yields the band edge.
Confidence ranges from –0.3 to 1; in units of 0.01. */

```
      VAR CONFIDENCE
            TYPE SIGNED BYTE
            MIN –30
            MAX 100

            /* Begin Confidence membership functions (VL, L, M, H, VH) */

            MEMBER VL
POINTS –30 0 –10 1 0 1 10 1 30 0
            END

            MEMBER L
POINTS 10 0 30 1 50 0
            END

            MEMBER M
POINTS 30 0 50 1 70 0
            END
```

```
             MEMBER H
POINTS 50 0 70 1 90 0
             END

             MEMBER VH
POINTS 70 0 90 1 100 1
             END

        END  /* End of Confidence definition */


        /* FUZZY RULE BASE */

        FUZZY L_EDGINESS_RULES

                RULE RULE1
IF (LEVEL IS ZO) THEN
        CONFIDENCE IS VL
END

                RULE RULE2
IF (SLOPE IS NBIG) THEN
        CONFIDENCE IS VL
END

                RULE RULE3
IF (LEVEL IS VH) THEN
        CONFIDENCE IS VL
END

                RULE RULE4
IF (LEVEL IS L) AND (SLOPE IS PMED) THEN
        CONFIDENCE IS M
END

                RULE RULE5
IF (LEVEL IS L) AND (SLOPE IS NMED) THEN
        CONFIDENCE IS VL
END

                RULE RULE6
IF (LEVEL IS L) AND (SLOPE IS PBIG) THEN
        CONFIDENCE IS L
END

                RULE RULE7
IF (LEVEL IS L) AND (SLOPE IS ZO) THEN
        CONFIDENCE IS VL
END
```

Chap. 11 Fuzzy Logic to Configure a Digital Filter   255

```
            RULE RULE8
IF (LEVEL IS M) AND (SLOPE IS PMED) THEN
     CONFIDENCE IS VH
END

            RULE RULE9
IF (LEVEL IS M) AND (SLOPE IS NMED) THEN
     CONFIDENCE IS L
END

            RULE RULE10
IF (LEVEL IS M) AND (SLOPE IS PBIG) THEN
     CONFIDENCE IS VL
END

            RULE RULE11
IF (LEVEL IS M) AND (SLOPE IS ZO) THEN
     CONFIDENCE IS L
END

            RULE RULE12
IF (LEVEL IS H) AND (SLOPE IS PMED) THEN
     CONFIDENCE IS L
END

            RULE RULE13
IF (LEVEL IS H) AND (SLOPE IS NMED) THEN
     CONFIDENCE IS L
END

RULE RULE14
IF (LEVEL IS H) AND (SLOPE IS PBIG) THEN
     CONFIDENCE IS VL
END

RULE RULE15
IF (LEVEL IS H) AND (SLOPE IS ZO) THEN
     CONFIDENCE IS VL
END

     END /* END OF KNOWLEDGE BASE */

     CONNECT
          FROM SLOPE
          TO L_EDGINESS_RULES
     END
```

```
CONNECT
     FROM LEVEL
     TO L_EDGINESS_RULES
END

CONNECT
     FROM L_EDGINESS_RULES
     TO CONFIDENCE
END

END /* Project Filt_Lo */
```

Chap. 11   Fuzzy Logic to Configure a Digital Filter   **257**

APPENDIX C
FUZZY-LOGIC RULEBASE FOR UPPER-EDGE DETECTION

PROJECT Filt_Hi

```
    /* SLOPE OF THE FILTER AT A GIVEN POINT.
    *  THE UNIVERSE OF DISCOURSE IS -10 TO 10.
    *  SLOPE IS MEASURED IN UNITS OF 0.1. */


    VAR SLOPE
        TYPE SIGNED BYTE
        MIN -100
        MAX 100

        /* membership fuctions for Slope (NBIG, NMED, ZO, PMED, PBIG). */

            MEMBER ZO
POINTS -5 0 0 1 5 0
            END

            MEMBER PMED
POINTS 0 0 5 1 30 1 40 0
            END

            MEMBER NMED
POINTS -40 0 -30 1 -5 1 0 0
            END

            MEMBER PBIG
POINTS 30 0 40 1 100 1
            END

            MEMBER NBIG
POINTS -100 1 -40 1 -30 0
            END


        END /* END OF SLOPE DEFINITION. */

    /* POWER LEVEL OF THE INPUT AT A GIVEN FREQUENCY,
        NORMALIZED WITH RESPECT TO THE MAXIMUM POWER.
        UNIVERSE OF DISCOURSE IS 0 TO 1; EXPRESSED IN TERMS OF
        0.005 NORMALIZED POWER UNITS. */
```

VAR LEVEL
    TYPE UNSIGNED BYTE
    MIN 0
    MAX 200

    /* Membership functions for LEVEL: ZO, VS, S, M, L, VL */

    MEMBER ZO
    POINTS 0 1 20 0
    END

    MEMBER L
    POINTS 0 0 20 1 40 0
    END

    MEMBER M
    POINTS 20 0 40 1 70 1 90 0
    END

    MEMBER H
    POINTS 70 0 90 1 110 0
    END

    MEMBER VH
    POINTS 90 0 110 1 200 1
    END

END /* End of LEVEL definition */

/* Confidence that a given point is an edge of the frequency band. The confidence vs. frequency curve forms the membership function for a fuzzy variable indicating the edge of the frequency band. Defuzzifying (by an external, non-Togai routine yields the band edge.
Confidence ranges from –0.3 to 1; in units of 0.01. */

VAR CONFIDENCE
    TYPE SIGNED BYTE
    MIN –30
    MAX 100

    /* Begin Confidence membership functions (VL, L, M, H, VH) */

    MEMBER VL
    POINTS –30 0 –10 1 0 1 10 1 30 0
    END

```
              MEMBER L
POINTS 10 0 30 1 50 0
              END

              MEMBER M
POINTS 30 0 50 1 70 0
              END

              MEMBER H
POINTS 50 0 70 1 90 0
              END

              MEMBER VH
POINTS 70 0 90 1 100 1
              END

      END  /* End of Confidence definition */


      /* FUZZY RULE BASE */

      FUZZY L_EDGINESS_RULES

              RULE RULE1
IF (LEVEL IS ZO) THEN
      CONFIDENCE IS VL
END

              RULE RULE2
IF (SLOPE IS PBIG) THEN
      CONFIDENCE IS VL
END

              RULE RULE3
IF (LEVEL IS VH) THEN
      CONFIDENCE IS VL
END

              RULE RULE4
IF (LEVEL IS L) AND (SLOPE IS NMED) THEN
      CONFIDENCE IS M
END

              RULE RULE5
IF (LEVEL IS L) AND (SLOPE IS PMED) THEN
      CONFIDENCE IS VL
END
```

```
        RULE RULE6
IF (LEVEL IS L) AND (SLOPE IS NBIG) THEN
      CONFIDENCE IS L
END


        RULE RULE7
IF (LEVEL IS L) AND (SLOPE IS ZO) THEN
      CONFIDENCE IS VL
END


        RULE RULE8
IF (LEVEL IS M) AND (SLOPE IS NMED) THEN
      CONFIDENCE IS VH
END


        RULE RULE9
IF (LEVEL IS M) AND (SLOPE IS PMED) THEN
      CONFIDENCE IS L
END


        RULE RULE10
IF (LEVEL IS M) AND (SLOPE IS NBIG) THEN
      CONFIDENCE IS VL
END


        RULE RULE11
IF (LEVEL IS M) AND (SLOPE IS ZO) THEN
      CONFIDENCE IS L
END


        RULE RULE12
IF (LEVEL IS H) AND (SLOPE IS NMED) THEN
      CONFIDENCE IS L
END


        RULE RULE13
IF (LEVEL IS H) AND (SLOPE IS PMED) THEN
      CONFIDENCE IS L
END


        RULE RULE14
IF (LEVEL IS H) AND (SLOPE IS NBIG) THEN
      CONFIDENCE IS VL
END


        RULE RULE15
IF (LEVEL IS H) AND (SLOPE IS ZO) THEN
      CONFIDENCE IS VL
END
```

END /* END OF KNOWLEDGE BASE */

CONNECT
    FROM SLOPE
    TO L_EDGINESS_RULES
END

CONNECT
    FROM LEVEL
    TO L_EDGINESS_RULES
END

CONNECT
    FROM L_EDGINESS_RULES
    TO CONFIDENCE
END

END /* Project Filt_Lo */

12

SIMULATION OF TRAFFIC FLOW AND CONTROL USING FUZZY AND CONVENTIONAL METHODS

Robert L. Kelsey
Keith R. Bisset
X-1, Los Alamos National Laboratory

12.1 INTRODUCTION

Fuzzy logic can be used as an alternative method for the control of traffic environments. A traffic environment includes the lanes to and from an intersection, the intersection, vehicle traffic, and signal lights in the intersection. To test the fuzzy logic controller a computer simulation was constructed to model a traffic environment. A typical cross intersection was chosen for the traffic environment and the performance of the fuzzy logic controller was compared with the performance of two different types of conventional control. The fuzzy logic controller proved to be a better method of control than conventional control methods especially in the case of highly uneven traffic flow between the different directions.

Chap. 12   Fuzzy Traffic Control   263

12.2   TRAFFIC CONTROL METHODS

There are two types of conventional control. One type of control uses a preset cycle time to change lights. The other type of control combines preset cycle times with proximity sensors which can activate a change in the cycle time or the lights. In the case of a less traveled street which may not need a regular cycle of green lights, a proximity sensor will activate a change in the light when cars are present. This type of control depends on having some prior knowledge of traffic flow patterns in the intersection so that signal cycle times and placement of proximity sensors may be customized for the intersection.

Fuzzy logic control is an alternative to conventional control which can control a wider array of traffic patterns at an intersection. A fuzzy-controlled signal light uses sensors that count cars instead of proximity sensors which only indicate the presence of cars. This provides the controller with traffic densities in the lanes and allows a better assessment of changing traffic patterns. As traffic distributions fluctuate, the fuzzy controller can change the signal light accordingly.

12.3   SIMULATION

The computer simulation was written in the C++ computer programming language using object-oriented programming techniques. Object-oriented programming allows the simulation to be built in a modular fashion making it both expandable and easily maintainable. Thus, new features are easily added to the simulation and current features can be modified with little impact on the rest of the program.

The graphics portion of the simulation was written using the X Window System as described by Nye (1988, 1991) and Young (1989), specifically the Xt Intrinsics library and the Athena (Xaw) and Hewlett-Packard (Xw) widget libraries, collectively called the X Toolkit. The use of the X Toolkit provides a set of widgets which can be combined together to create the simulation. A widget is a graphical object such as a button or a slider. Their use allows the low-level graphical details to be ignored.

The X Toolkit (and X Windows in general) is event driven. The simulation responds to events, such as a button press, generated by the user. Any one of a number of actions can be performed at any time which greatly enhances the interactive nature of the simulation.

12.3.1   Vehicle Movement

The crux of the simulation problem is the modeling of traffic flow to mimic reality, specifically the motion of the vehicles relative to one another. Physical equations were used to describe the motion of a car based on the car immediately in front of it (the leading car). Equation 1 describes the acceleration of a car based on the velocities and positions of the car and the leading car. This equation is a time-delay differential equation which comes from traffic flow theory presented by Haight (1963), Montrol (1961), and Morris (1967). Equations 2 and 3 are standard classical physics equations for the velocity and position of an object based on the object's acceleration. $v(t)$ and $X(t)$ are the velocity and position, respectively, of the car at time t, while $v'(t)$ and $x'(t)$ are the velocity and position of the leading car. τ is the driver reaction time and V_{free} is the free velocity of the car (the velocity at which the car would drive without interference from other cars).

$$a(t + \tau) = 4v(t) \frac{(v(t) - v'(t))}{(x(t) - x'(t)')^2} + 0.2 \, (V_{free} - v(t)) \tag{1}$$

$$v(t) = v(t-1) + a(t)dt \tag{2}$$

$$x(t) = x(t-1) + v(t)dt + \frac{1}{2}a(t)dt^2 \tag{3}$$

The free velocity of each car is randomly set from a normal distribution with a mean of the "posted" speed limit in the lane. As observed in reality, this provides the simulation with cars which drive the speed limit and cars which drive faster and slower than the speed limit. The driver reaction time, τ, is the time expended when a driver sees a disturbance, the brain processes the information, and the driver takes an appropriate action. This time is equal to one and one half seconds.

12.3.2   Input

The simulation has a variety of modes and other configuration inputs which may be selected from an input file when the program is executed. These inputs include the type of controller (timed, proximity, or fuzzy) which may be used to control the traffic signal, whether to use graphics or batch mode, flow rates, percentage of turning cars, and how long (in simulation time) the simulation should run.

Flow rate in cars per hour may be specified separately for each of the four lanes. This parameter may be either a single value (cars per hour) or a series of flow-time pairs (two numbers: flow and time). If a single value is given then that flow rate is used for the entire simulation run. The cars are actually generated and entered into the simulation randomly to obtain the specified flow rate, but to make the flow pattern or distribution of cars more nonuniform. The flow-time pairs give the flow rate at certain times during the simulation run. The simulation interpolates between these pairs to provide a smooth transition between flows. Flow-time pair inputs are useful for determining how the controller will perform in a given intersection over the course of an entire day, or some other time period. A sample input file that uses both flow-time pairs and single values is shown in Appendix A.

12.3.3   Operating Mode

The simulation has two operating modes. A graphics mode displays a scaled traffic environment and a batch mode executes without graphics. The graphics mode animates the movement of cars in lanes directed to and from the intersection. Also displayed are statistics about the cars which have moved through the traffic environment. These include flow rate (cars per hour), average waiting time, and average driving time. The average waiting time in seconds is an average of the time all cars spend at zero velocity in the simulation. The average driving time is the average of the time all cars (completing the

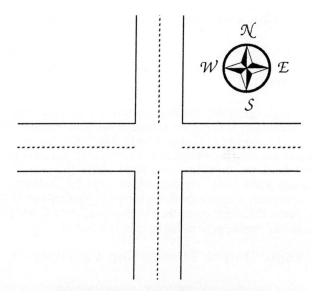

Figure 12-1. Cross Intersection.

simulation) spend at a velocity greater than zero during the execution of the simulation. In addition, the fuzzy controller version of the graphics simulation has "pop up" windows available which display the degree of membership (refer to Section 12.4.1) for each of the inputs, which fuzzy rules are being fired, and a list of the fuzzy rules. This information gives insight into the workings of the fuzzy controller.

      The graphics mode also allows certain inputs originally set in the input file to be changed interactively. These inputs include the flow rate in each direction and the cycle time length of the lights. Due to the computational overhead of generating graphics, the batch mode can execute as much as ten times faster than the graphics mode. This makes the batch mode ideal for running exhaustive tests using different traffic patterns for the three different controllers.

12.3.4 Evaluation Function

The statistics generated by the simulation are used to calculate a cost function (equation 4) as a means of evaluating the performance of the fuzzy controller against the conventional controllers. $Wait_{mean}$ is the average waiting time. $Drive_{mean}$ is the average drive time. $Cars_{out}$ and $Cars_{in}$ are the number of cars exiting and entering the traffic environment, respectively, while k is a constant which is equal to 100. The lower the cost function, the better the performance.

$$Cost = \frac{Wait_{mean}}{k \left(\frac{Cars_{out}}{Cars_{in}} \right) Drive_{mean}} \qquad (4)$$

12.4 FUZZY CONTROLLER

A fuzzy logic controller was designed for a typical cross intersection traffic environment. The cross intersection has a lane directed to and from the intersection at each compass point as shown in Figure 12-1.

The fuzzy logic controller was implemented using Togai InfraLogic's Fuzzy C Compiler (1991). The Fuzzy C Compiler accepts a fuzzy source code file as input and compiles this file into a module of C source code. The fuzzy source code is much like natural language that describes each input to and output from the fuzzy controller and the fuzzy membership functions associated with those inputs and outputs. Also described with natural language is the fuzzy rule base which maps the combinations of inputs to the outputs. A sample fuzzy source code file is shown in Appendix B. The C source code generated by the Fuzzy C Compiler can then be compiled (by a C compiler) within a C program or as a separate routine to be called by other programs.

12.4.1 Input/Output Membership Functions

There are three inputs into the fuzzy controller, the average density of traffic behind the green lights, the average density of traffic behind the red lights, and the length of the current cycle time. There are four membership functions describing the densities of traffic at the green and red lights. These functions are labeled Zero, Low, Medium, and High, and are shown in Figures 12-2 and 12-3. The membership functions for the green and red densities need to be different due to the fact that the densities of the cars in those two cases will be different. Cars stopped at a red light are much closer together than cars moving through a green light, therefore the density of the cars at the red light will be higher than that of the cars moving through the green light. This detail was actually discovered while observing the graphic display of the simulation. The first version of the fuzzy logic controller used the same membership functions for green and red densities. The third input, current cycle time, has three membership functions describing time in seconds. These functions are Short, Medium, and Long, and are shown in Figure 12-4.

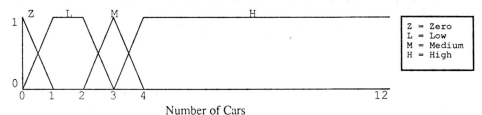

Figure 12-2 Membership functions for green lights.

The densities are taken from two sensors placed on the road. One is at the intersection and the other is 150 feet before the light. The rear sensor increments a counter every time a car passes over it while the forward sensor decrements the same counter. In this way a count of the number of cars in the 150 feet before the light is obtained. This is in contrast to conventional control which places a proximity sensor at the light and can only sense the presence of cars waiting at a light, not how many cars are waiting.

Chap. 12 Fuzzy Traffic Control

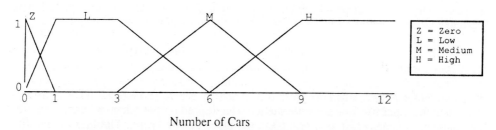

Figure 12-3. Membership functions for red lights.

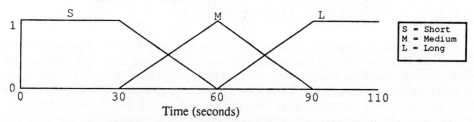

Figure 12-4. Membership functions for time.

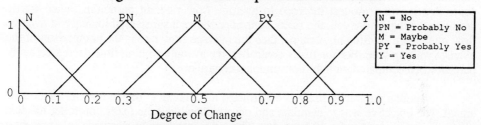

Figure 12-5. Membership functions for change.

The output of the fuzzy controller decides whether to change the state of the light (for example, green north-south to green east-west) or remain in the same state. The membership functions for change are No, Probably No, Maybe, Probably Yes, and Yes. These functions represent a degree (fuzzy value) of a binary value, 1 being yes and 0 being no, and are shown in Figure 12-5.

Other attempts at fuzzy control of traffic signals, particularly by Gallegos (1991), have used the fuzzy controller to make incremental changes in the times of the cycle in each direction. We rejected that approach because it was felt that a controller which could make immediate changes to the light based on the current conditions was preferable.

12.4.2 Fuzzy Rule Base

The fuzzy rule base maps the combination of the inputs to the output to decide whether to change the light. In the fuzzy source code file the fuzzy rule base is a list of if/then statements. The number of rules is equal to the number of input combinations derived from the number of membership functions per input. For example, if there are two inputs each having three membership functions, then the number of rules would equal nine.

It is possible to scale down the number of fuzzy rules when certain input combinations are unnecessary. The fuzzy traffic controller design presented would yield a rule base of 48 rules, but for certain combinations the input for cycle time has no meaning. For example, when the green traffic density is zero and the red traffic density is zero then the input for time does not matter. The desired output (change) is the same for any amount of time. Thus, one rule takes the place of three rules. The fuzzy controller presented uses 26 different fuzzy rules as shown in Appendix C.

12.5 TESTING AND RESULTS

Testing of the fuzzy controller involved the simulation of 625 different parameter sets. These parameter sets vary the flow of traffic in each of four lanes, from 0 to 1200 cars per hour, with an increment of 300 (hence, 625 combinations). Each of these parameter sets is simulated for one hour of simulation time and generate statistics including the average flow rate (cars/hour), average driving time, and average waiting time. Results, in the form of the statistics, of the fuzzy controller simulations are compared with the same 625 parameter sets using a conventional controller (cycle time controller). Overall, the fuzzy controller shows a modest increase in average flow rate. In particular, when traffic densities are highly uneven, the fuzzy controller shows a substantial increase in average flow rate. More significant, perhaps, then the average flow rate is the comparison of the average waiting time. For the fuzzy controller, the average waiting time decreased by 49 percent when compared with the conventional controller.

Another parameter file was constructed to simulate a "day in the life" of a traffic signal. The objective was to simulate and observe how well the fuzzy controller handled traffic during the span of a 12 hour day. The cross intersection was used to represent a typical city intersection with the north-south direction being a main artery and the east-west direction being a feeder artery. The 12 hour day included morning rush time, morning lull, lunch rush time, afternoon lull, school's out rush time, and evening rush time. Again, the fuzzy-controller simulation was compared with conventional-controller simulation. The fuzzy controller handled ten percent more cars than the conventional controller. The fuzzy controller also showed an eight percent increase in average flow rate, a 64 percent decrease in average waiting time, and a 29 percent decrease in the maximum waiting time. Figure 12-6 shows the flow rates throughout the 12 hour day. North-south and east-west flow rates are represented. Figure 12-7 shows a performance comparison of the fuzzy controller with two different conventional controllers across the 12 hour simulation. One conventional controller is the cycle time controller and the other is a cycle time controller enhanced with proximity sensors (refer to Section 12-5.2 Additional Testing).

The 625 different combinations simulations were useful in identifying strengths and weaknesses in the fuzzy controller. Using this information, we were able to modify aspects of the fuzzy controller and improve its performance. The "day in the life" simulation and the corresponding results are more interesting because they provide a "real

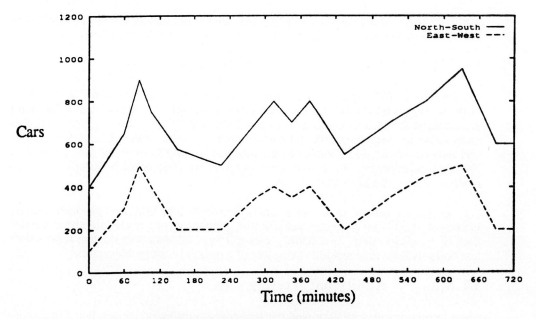

Figure 12-6. 12 hour flow rates.

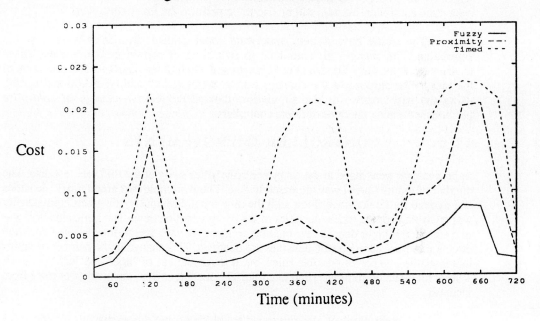

Figure 12-7. Controller performance comparison.

269

world" example. This example could be easily applied and modified to simulate an actual intersection in a city.

12.5.1 Streamlined Test Suite

The test suite of 625 parameter sets was eventually collapsed to 69 sets due to the nature of the cross intersection and the duplicated traffic patterns in each lane. Since the inputs are the same for each of the four lanes and the traffic environment is symmetrical it is unnecessary to repeat the traffic patterns for each lane. The results of the 625 parameter sets showed little to no difference in repeated traffic patterns from one lane to another. This is to be expected. What little difference was observed could be explained by the random nature of the input traffic flows.

The 69 parameter sets vary traffic flow from 0 to 1400 cars per hour with an increment of 350 cars per hour. Each of these parameter sets is also simulated for one hour of simulation time. This scaled down test suite allows for evaluation results with less computational time and little or no loss of coverage of the problem space.

12.5.2 Additional Testing

Later testing included an improved conventional controller which combined proximity sensors with preset cycle times. This controller performed much better than the preset cycle time controller, but was still unable to outperform the fuzzy controller.

The traffic environment underwent some changes in a later version of the simulation. The changes all amounted to re-scaling of certain measurements. These measurements included the size of cars, length and width of lanes, and sensor placement in the lanes. The purpose of the changes was to better reflect reality. Additional testing using the new version of the simulation showed no relative change in controller performance among the three different controllers.

12.6 CONTROLLER IMPROVEMENTS

Improvements were made to the fuzzy controller after study of the different test data. The number of inputs (three) was increased to five. There are now two green traffic densities and two red traffic densities along with the time input. Previously, the green input density was an average of the traffic densities in the two green directions. The same situation was true for the red input density. By making an input for each green direction and each red direction more accurate traffic densities are obtained. This change in the number of inputs increases the number of possible rules, but through the use of "and" and "or" operators, several rules can be combined into one rule. This technique kept the number of rules from increasing.

The controller was also improved by making some minor changes to the shapes of some of the membership functions in the controller inputs. These changes were made after careful study of the test results and observation of the graphic display of the simulation. Each change to a membership function was followed by additional testing to verify performance.

12.7 CONCLUSIONS

Preliminary results show that larger quantities of traffic are "handled" by fuzzy control methods then by conventional control methods. A "handled" car is one that has exited the simulation after having passed through the intersection. Fuzzy control shows greatest improvement over conventional methods when the traffic flow is highly uneven. The fuzzy controller can change the lights as necessary to achieve the maximum throughput, rather than be limited to a preset cycle time for changing signal lights. Another, perhaps more important result, is that the average time spent waiting in traffic decreases with the use of fuzzy control versus conventional control. This means individual cars spend less time waiting and more time moving.

There are other benefits to fuzzy control as well. The fuzzy controller handles a wide range of continually changing traffic patterns. This characteristic makes the fuzzy controller reusable, an aspect not shared by conventional controllers. This concept, even with modest or no improvement in controller performance, makes a fuzzy controller more desirable. For example, all cross intersections may have the same fuzzy controller without the need for individual controller configuration based on observations at an intersection. Also, the cost of a traffic-control system may actually decrease due to reusability.

12.8 FUTURE WORK

For the future we would like to improve the performance of the fuzzy controller by adjusting the membership functions of the inputs. The number of points in the membership functions that might be moved and the number of inputs to the controller yield a large number of possible combinations which could be inspected. Although it would be interesting to test each combination with the suite of 69 parameter tests, the amount of computational time involved prevents this course of action. Instead, we intend to use a neural network that can produce a surface representing all the combinations from a few of the tested combinations. From this surface we hope to pick a combination of points which will yield a better performing fuzzy controller.

Although this would not be a real-time adaptive controller, it might suggest whether a real-time adaptive controller is possible and of interest. The fuzzy controller provides a "global" solution to the problem space in that the fuzzy controller handles a wide range of traffic patterns. By adapting the fuzzy controller with a neural network a better "global" solution is realized. Constant (real time) adaptation of the fuzzy controller may not be necessary.

12.9 ACKNOWLEDGMENTS

The authors wish to thank groups EES-13 and X-1 at Los Alamos National Laboratory for their support in this project and Donn Hines for his valuable suggestions.

REFERENCES

Bisset, K. and R. Kelsey, "Simulation of Traffic Flow and Control Using Conventional, Fuzzy and Adaptive Methods, *Proc. of the 1992 European Simulation Multiconference*, June 1992.

Gallegos, R. and T. Nguyen, "Fuzzy Logic Traffic Application", EECE 446 Technical Report, University of New Mexico, Albuquerque, NM. 1991.

Haight, Frank A., *Mathematical Theories of Traffic Flow*, New York: Academic Press Inc. 1963.

Montrol, E. W., "Acceleration Noise and Clustering Tendency of Vehicular Traffic", *Theory of Traffic Flow*, ed. R. Herman, New York: Elsevier Publishing Company, 1961.

Morris, R. W. J. and P. G. Pak-Poy, "Intersection Control by Vehicle Actuated Signals", *Vehicular Traffic Science*, ed. R. Herman and R. Rothery, New York: American Elsevier Publishing Company, Inc. 1967.

Nye, A., *Xlib Programming Manual*. Sebastopol, CA: O'Reilly and Associates, Inc., 1988.

Nye, A. and T. O'Reilly, X *Toolkit Intrinsics Programming Manual*. Sebastopol, CA: O'Reilly and Associates, Inc., 1990.

Togai InfraLogic, *Professional Fuzzy-C Development System User's Manual*. Irvine, CA: Togai Infralogic, Inc., 1991.

Young, D. A., X *Window Systems: Programming and Applications With Xt*, Englewood Cliffs, New Jersey: Prentice-Hall, Inc., 1989.

APPENDIX A

fuzzy
graphic display on
simulation time 3:00:00

Flow Rate
    North 500 at 0, 700 at 30:00, 1000 at 1:00:00, 800 at 1:30:00, 700 at 1:45:00
    South 600 at 0, 500 at 30:00, 900 at 1:00:00, 800 at 1:30:00, 500 at 1:45:00
    East 500 at 0, 700 at 30:00, 1000 at 45:00, 800 at 1:45:00, 700 at 2:00:00
    West 950

Turn rate
    north right 15 at 0, 10 at 45:00, 15 at 1:00
    north left 5
    south right 15
    south left 5
    east right 15
    east left 5
    west right 15
    west left 5

APPENDIX B

```
PROJECT Traf 1
VAR green
    TYPE float
        MIN 0
        MAX 12

    MEMBER Z
        POINTS 0 1 1 0
    END

    MEMBER L
        POINTS 0 0 1 1 2 1 3 0
    END

    MEMBER M
        POINTS 2 0 3 1 4 0
    END

    MEMBER H
        POINTS 3 0 4 1 1 2 1
    END

END
```

VAR red
    TYPE float
    MIN 0
    MAX 12

    MEMBER Z
        POINTS 0 1 1 0
    END

    MEMBER L
        POINTS 0 0 1 1 3 1 6 0
    END

    MEMBER M
        POINTS 3 0 6 1 9 0
    END

    MEMBER H
        POINTS 6 0 9 1 12 1
    END

END

VAR time
    TYPE float
    MIN 8
    MAX 110

    MEMBER S
        POINTS 8 1 30 1 60 0
    END

    MEMBER M
        POINTS 30 0 60 1 90 0
    END

    MEMBER L
        POINTS 60 0 90 1 110 1
    END

END

VAR change
    TYPE float
    MIN 0
    MAX 1

Chap. 12 Fuzzy Traffic Control

```
    MEMBER N
        POINTS 0 1 .2 0
    END

    MEMBER PN
        POINTS .1 0 .3 1 .5 0
    END

    MEMBER M
        POINTS .3 0 .5 1 .7 0
    END

    MEMBER PY
        POINTS .5 0 .7 1 .9 0
    END

    MEMBER Y
        POINTS .8 0 1 1
    END

END


FUZZY traf 1_control
    RULE Rule_0
        IF green is Z and red is Z THEN change is N
    END

    RULE Rule_1
        IF green is Z and red is L THEN change is Y
    END

    RULE Rule_2
        IF green is Z and red is M THEN change is Y
    END

    RULE Rule_3
        IF green is Z and red is H THEN change is Y
    END

    RULE Rule_4
        IF red is Z THEN change is N
    END

    RULE Rule_5
        IF green is L and red is L THEN change is N
    END
```

```
RULE Rule_6
    IF green is M and red is M THEN change is N
END

RULE Rule_7
    IF green is H and red is H THEN change is N
END

RULE Rule_8
    IF green is L and red is M and time is S THEN change is M
END

RULE Rule_9
    IF green is L and red is M and time is M THEN change is PY
END

RULE Rule_10
    IF green is L and red is M and time is L THEN change is Y
END

RULE Rule_11
    IF green is L and red is H and time is S THEN change is PN
END

RULE Rule_12
    IF green is L and red is H and time is M THEN change is M
END

RULE Rule_13
    IF green is L and red is H and time is L THEN change is PY
END

RULE Rule_14
    IF green is M and red is L and time is S THEN change is PN
END

RULE Rule_15
    IF green is M and red is L and time is M THEN change is PN
END

RULE Rule_16
    IF green is M and red is L and time is L THEN change is M
END

RULE Rule_17
    IF green is M and red is H and time is S THEN change is M
END
```

RULE Rule_18
    IF green is M and red is H and time is M THEN change is PY
END

RULE Rule_19
    IF green is M and red is H and time is L THEN change is Y
END

RULE Rule_20
    IF green is H and red is L and time is S THEN change is M
END

RULE Rule_21
    IF green is H and red is L and time is M THEN change is PY
END

RULE Rule_22
    IF green is H and red is L and time is L THEN change is Y
END

RULE Rule_23
    IF green is H and red is M and time is S THEN change is PN
END

RULE Rule_24
    IF green is H and red is M and time is M THEN change is PN
END

RULE Rule_25
    IF green is H and red is M and time is L THEN change is M
END

END

CONNECT
    FROM green to traf 1_control
END

CONNECT
    FROM red to traf 1_control
END

CONNECT
    FROM time to traf 1_control
END

CONNECT
    FROM traf 1_control to change
END

APPENDIX C

1. IF green is zero and red is zero THEN change is no
2. IF green is zero and red is low THEN change is yes
3. IF green is zero and red is medium THEN change is yes
4. IF green is zero and red is high THEN change is yes
5. IF red is zero THEN change is no
6. IF green is low and red is low THEN change is no
7. IF green is medium and red is medium THEN change is no
8. IF green is high and red is high THEN change is no
9. IF green is low and red is medium and time is short THEN change is maybe
10. IF green is low and red is medium and time is medium THEN change is probably yes
11. IF green is low and red is medium and time is long THEN change is yes
12. IF green is low and red is high and time is short THEN change is probably no
13. IF green is low and red is high and time is medium THEN change is maybe
14. IF green is low and red is high and time is long THEN change is probably yes
15. IF green is medium and red is low and time is short THEN change is probably no
16. IF green is medium and red is low and time is medium THEN change is probably no
17. IF green is medium and red is low and time is long THEN change is maybe
18. IF green is medium and red is high and time is short THEN change is maybe
19. IF green is medium and red is high and time is medium THEN change is probably yes
20. IF green is medium and red is high and time is long THEN change is yes
21. IF green is high and red is low and time is short THEN change is maybe
22. IF green is high and red is low and time is medium THEN change is probably yes
23. IF green is high and red is low and time is long THEN change is yes
24. IF green is high and red is medium and time is short THEN change is probably no
25. IF green is high and red is medium and time is medium THEN change is probably no
26. IF green is high and red is medium and time is long THEN change is maybe

13

A FUZZY GEOMETRIC PATTERN RECOGNITION METHOD WITH LEARNING CAPABILITY

Scott Peterson
Clin Lashway
Raydec Inc.

Doug Miller
University of New Mexico

This chapter describes a fuzzy-logic approach to pattern recognition of geometric shapes. Also utilized here are genetic algorithms. The results show great deal of promise with a strong proof of principle for pattern recognition problems.

13.1 INTRODUCTION

This chapter describes a method of geometric pattern recognition that incorporates elements of fuzzy logic and genetic algorithms. The recognition process is implemented using fuzzy measure to determine pattern characteristic class membership and a genetic algorithm to adaptively establish classification parameters. There were two overall objectives in the development of this procedure. First, we wish to show that by using a fuzzy measure it is possible to construct a recognition process that requires a relatively small amount of pattern information. Second, we wish to establish a proof-of-concept link between a fuzzy classification scheme and a learning procedure based on genetic algorithms. To achieve this objective, pattern recognition was restricted to four simple geometric shapes (circle, square, rectangle, and rhombus) and only three pattern grid measurements were chosen to be made (pattern centroid, maximum polar coordinates, and centroid axis intercept points). The actual recognition process consists of classifying any given input pattern into one of the four fuzzy pattern sets. Pattern classification is based on a linear combination of fuzzy characteristic measures. During a training cycle, known patterns are read from data files and the linear parameters are continuously generated and evaluated until a desired performance level is obtained.

Although initial results are currently under analysis, the system has shown promise in that it obtains learning stage performance levels of 96 to 100 percent for relatively small training files (10 to 30) patterns and obtains 85 to 90 percent levels for larger training files containing a number of distorted figures.

The pattern recognition algorithm consists of two main parts: The fuzzy recognition procedure and the genetic algorithm. Section 13.2 describes the development of the fuzzy recognition procedure. Section 13.3 contains a brief introduction to the genetic algorithm as they pertain to this procedure and details its implementation for this application. Initial results are given in section 13.4 along with a description further work to be carried out on the project.

13.2 RECOGNITION PROCEDURE

Input to the recognition algorithm consists of a 100 by 100 binary grid. So as to keep software development within the scope of our main objectives, we assume that each input pattern is one of the four figures under consideration, all pattern lines are connected, and the figure is correctly oriented within the grid (that is, the pattern has not been rotated). A typical pattern is shown in Figure 13-1.

The fuzzy recognition process is shown in the block diagram in Figure 13-2 and consists of two main portions; a pattern characteristic extractor and a fuzzy classifier. These processes are described further in the following subsections.

Pattern Characteristic Extraction

The characteristic extractor scans the input grid to compute the pattern centroid. An x-y coordinate system is then established relative to the centroid position and the x- and y-axis

Chap. 13 Fuzzy Geometric Pattern Recognition

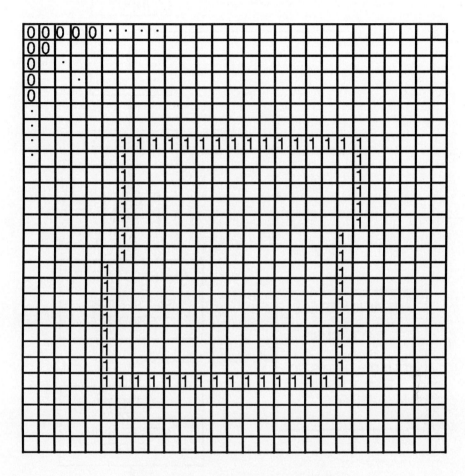

Figure 13-1. Pattern Input Grid.

intercepts are determined. Next the maximum radius and polar angle in all four quadrants are computed. Finally, the mean and standard deviation of the pattern relative to its centroid are computed and stored. Characteristics of a typical input pattern are shown in Figure 13-3.

Fuzzy Classifier

The fuzzy classifier system determines the membership grade of the input pattern for each of the fuzzy pattern sets. This process begins with a series of fuzzy characteristic measures based on a combination of the axis intercepts, maximum radii and pattern statistics mentioned in the previous section. These fuzzy measures are computed from the following set of membership grades:

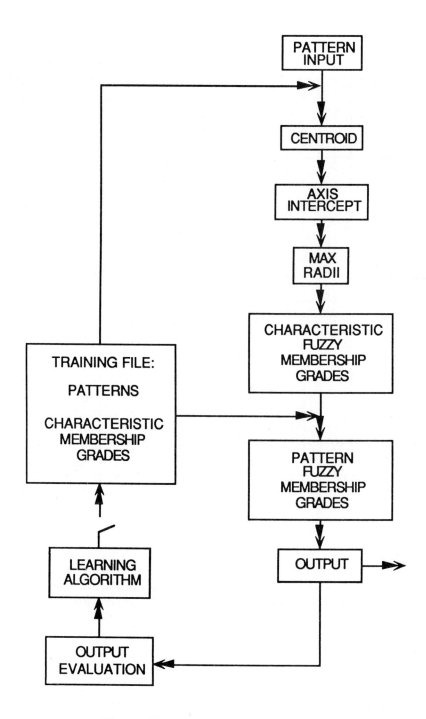

Figure 13-2. Recognition System.

Chap. 13   Fuzzy Geometric Pattern Recognition

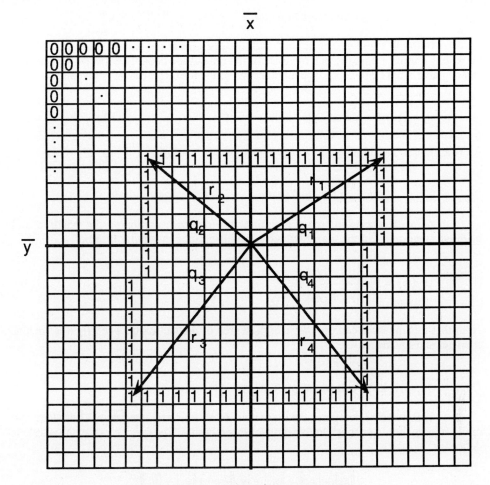

Figure 13-3. Pattern Characteristics.

1. Origin Symmetry

The symmetry of the maximum pattern radii about the origin helps provide distinction between figures such as squares and triangles. The measure compares the polar angles of the maximum radii in all four quadrants and the membership grade is given by

$$\mu_s(p) = 1 - \frac{1}{\pi}(|\theta_1 - \theta_3| + |\theta_2 - \theta_4|) \tag{1}$$

Note that this measure in indeterminate for circles since ideally there should be no maximum radius.

2. X-Axis Intercept Equality

With the x and y axes defined with their origins at the pattern centroid position, the x-axis intercept equality is a measure of the magnitude equality between the positive and negative x-axis intercepts. The membership grade for this measure is given by

$$\mu_x(p) = 1 - \frac{||x_1| - |x_2||}{|x_1| + |x_2|} \qquad (2)$$

All four figures should maintain a high degree of this measure.

3. Y-Axis Intercept Equality

As with x-axis intercept equality, this measure compares the positive and negative magnitudes of the y-axis intercepts and is defined in similar fashion.

$$\mu_y(p) = 1 - \frac{||y_1| - |y_2||}{|y_1| + |y_2|} \qquad (3)$$

Note that this measure can provide a high degree of distinction between figures such as triangles and squares.

4. X-Y Axis Intercept Equality

This measure compares the magnitude of all four axis intercepts. The square and circle figures should posses a high degree of this measure. The fuzzy membership grade is defined as

$$\mu_{xy}(p) = 1 - \frac{||x_1| - |x_2|| + ||x_1| - |y_1|| + \cdots + ||y_1| - |y_2||}{|x_1| + |x_2| + |y_1| + |y_2|} \qquad (4)$$

5. Maximum Radii Equality

The maximum radii equality measure provides discrimination between skewed figures such as the rhombus and other figures. It compares the maximum radii in all four quadrants according to the fuzzy membership grade

$$\mu_r(p) = 1 - \frac{|r_1 - r_2| + |r_1 - r_3| + \cdots + |r_3 - r_4|}{r_1 + r_2 + r_3 + r_4} \qquad (5)$$

Chap. 13   Fuzzy Geometric Pattern Recognition

6.   Figure Aspect

This is a direct measure of figure aspect; length divided by height. While x-y axis intercept yields a similar measure, the purpose here is to provide a more dynamic means of rectangular discrimination. The membership grade is defined as

$$\mu_a(p) = 1 - \frac{|x_1| + |x_2|}{|y_1| + |y_2|} \tag{6}$$

7.   Mean Equality

Mean equality compares the mean of the input pattern to the mean of an ideal figure given the input pattern characteristics. For example, in computing the mean for an ideal rhombus, figure length and height would be estimated from axis intercept data and then the skew angle of the rhombus would be estimated using maximum radius data. Given these dimensions the statistical mean of the ideal figure can be computed and compared to the actual pattern mean. The similarity between the actual and pattern mean is measured using the fuzzy membership grade defined as

$$\mu_{\bar{r}}(p) = 1 - \frac{|\bar{r} - \bar{r}_p|}{k} \tag{7}$$

where k is a constant depending on the maximum expected mean value.

8.   Standard Deviation Equality

This measure is entirely similar to that of the mean. The membership grade is

$$\mu_\sigma(p) = 1 - \frac{|\sigma - \sigma_p|}{k} \tag{8}$$

The two remaining characteristic measures are the complements of aspect and maximum radius equality:

$$\begin{aligned}\mu_{\bar{R}}(p) &= 1 - \mu_R(p) \\ \mu_{\bar{a}}(p) &= 1 - \mu_a(p)\end{aligned} \tag{9}$$

The pattern membership grades can now be defined as the normalized weighted sum of the characteristic measures

$$\mu(p) = \frac{A_1\mu_s(p) + A_2\mu_x(p) + \cdots + A_{10}\mu_\sigma(p)}{A_1 + A_2 + A_3 + \cdots + A_{10}} \tag{10}$$

where $A_1, A_2, ..., A_{10}$ are constants to be determined. At this point it should be noted that the fuzzy recognition system is complete upon specification of the constants A_i. One method of determining these constants would be to manually inspect the outcome of several trial patterns resulting in a knowledge base from which the constants can be chosen. The advantage to this approach lies in the fact that there may be fuzzy measures that do not apply to certain pattern sets such as the case of origin symmetry for the circle figure. In this case one can simply corresponding coefficient to zero whereas any optimization method would yield a nonzero coefficient regardless of the data. The disadvantage of the manual inspection process however, is obvious especially when more complex recognition systems are considered.

13.3 THE GENETIC ALGORITHM

As previously mentioned, the function recognition system is complete upon the specification of the linear constants A_i in the pattern membership grade function. It is clear that these coefficients should be chosen such that they tend to maximize the number of correct identifications for a given set of test patterns. Due to the number of parameters and the variability of pattern conditions, a deterministic approach to coefficient selection is not readily apparent. We therefore consider a nondeterministic optimization process known as the genetic algorithm.

The central theme in any optimization process is to search or compute extremal points over some given domain. Genetic algorithms were developed by Holland [1] at the University of Michigan and are defined as search algorithms based on the mechanics of natural selection and natural genetics. One of the most important goals in researching genetic algorithms is the ability to design artificial systems software that retains the important mechanisms of natural systems thus leading to optimization processes that are generally more robust in dealing with information which is vast and less certain [2]. This point suggests that the genetic algorithm is well suited for use with fuzzy logic systems.

Population Search

A population is defined as a current sample of the search domain. In this case the population consists of vectors whose components are the linear pattern membership function parameters:

$$\begin{matrix} [a_0 & a_1 & \cdots & a_n] \\ [b_0 & b_1 & \cdots & b_n] \\ & \vdots & & \\ [c_0 & c_1 & \cdots & c_n] \end{matrix} \tag{11}$$

The genetic algorithm under consideration here consists of three operations on a randomly generated population. These operations are reproduction, cross-over and mutation.

Chap. 13 Fuzzy Geometric Pattern Recognition

Pairs of individuals from the population are chosen at random to reproduce. The probability that any given individual is chosen for reproduction is proportional to that individual's "fitness" level. This fitness levels defined as the mathematical value of the objective function at the point in the domain represented by the individual. Hence, the higher the fitness level, the closer the individual is to the extreme point. In this case the objective function is simply the number of test patterns correctly identified by the recognition process given the set of linear coefficients contained in an individual of the population.

After all pairs of individuals have been selected from the population, crossover reproduction occurs. During this process a crossover point is uniform randomly generated for each pair and two new children are produced by swapping the high end of the parent vectors as shown in Figure 13-4.

$$\left\{ \begin{matrix} [a_0 & a_1 & a_2 & \cdots & a_c & a_{c+1} & \cdots & a_{n-1} & a_n] \\ [b_0 & b_1 & b_2 & \cdots & b_c & b_{c+1} & \cdots & b_{n-1} & b_n] \end{matrix} \right\}$$

$$\Downarrow$$

$$\left\{ \begin{matrix} [a_0 & a_1 & a_2 & \cdots & a_c & b_{c+1} & \cdots & b_{n-1} & b_n] \\ [b_0 & b_1 & b_2 & \cdots & b_c & a_{c+1} & \cdots & a_{n-1} & a_n] \end{matrix} \right\}$$

Figure 13-4. Crossover reproduction.

The children replace their parents in the population and the process is repeated with the next generation. The idea behind the reproduction of generations of the population is that coefficient vectors that yield the best results will tend to be more likely candidates for reproduction. Since the parent vectors pass on some of their characteristics to their children it is reasonable to assume that the next generation is likely to develop yet more favorable characteristics.

It is possible that the process may suffer early convergence to local extrema if there is no mechanism to randomly inject new possibility into the population (a somewhat lose analogy can be drawn here to the biological phenomenon of inbreeding). To prevent this early convergence the component of an individual vector component is randomly selected and changed. This process occurs according to a preset probability defined as the mutation rate. Vector component mutation helps assure that the process will not limit itself to an exclusive subset of coefficients that may not be optimal. Of course as the process begins to converge on the optimal vector population mutation can become detrimental. To achieve a stable steady-state the mutation rate is decreased as the average strength of the population increases.

13.4 RESULTS

Both the recognition process and the genetic algorithm were implemented in the C programming language. Initial testing has been performed using linear coefficients derived both by inspection and using the genetic training procedure. The following subsections describe the results obtained using these two methods.

Linear Coefficient by Inspection

During this portion of testing, linear coefficients were chosen to be either zero or one. The zeroed coefficients were those corresponding to characteristic measures perceived as having little or no relevance to the pattern membership grade under consideration. For example, an ideal circle should have a constant radius from the centroid origin. Therefore, a measure of origin symmetry is of little value and is set to zero. On the other hand, a measure of figure aspect is very relevant to rectangle discrimination and the coefficient corresponding to aspect is set to unity. The overall performance of the recognition procedure using this method was about fifty percent correct identification with most discrimination difficulty between squares and rectangles.

Linear Coefficients by Genetic Algorithm Training

Testing of the genetic algorithm portion of the system is performed by generating a series of test patterns which are submitted to the recognition part of the system resulting in an array of characteristic measures. These measures are stored in a test data file and read into a stand-alone version of the genetic training system. To evaluate the performance of the system, the number of correctly identified patterns at each generation of the population are written to a data file.

The algorithm has currently been subjected to three main test cases. The first case consisted of twelve perfect figures; three circles, three squares, three rectangles, and three rhombuses. The algorithm achieved 100 percent accuracy in this case within twenty generations.

The second test case contained twenty eight patterns. All of the first case patterns were used and patterns representing slightly stretched circles and squares were added along with one very elongated rhombus. The results of this test are shown by the graph in Figure 13-4. From the graph in Figure 13-4 it is apparent that the algorithm achieves about a 95 percent accuracy rate for the best coefficient vector with two hundred fifty generations. The population average climbs to this rate within three hundred generations. These results imply that linear coefficients have been chosen such that the recognition system should correctly identify twenty seven out of twenty eight nearly perfect figures. Table 13-1 shows the resulting pattern membership grades upon algorithm convergence. The columns of this table represent the pattern membership grades computed for each of the four figures under consideration in this test. The rows represent the results of each test pattern. It is apparent from this table that the highest membership grades do indeed correspond to the correct figure except for test pattern number twenty eight. This case corresponds to the extremely elongated rhombus which was incorrectly identified as a

Chap. 13 Fuzzy Geometric Pattern Recognition   289

rectangle. It is also apparent that there is not much variation among the computed pattern membership grades resulting in less certain figure discrimination.

The lack of discrimination certainly manifests itself to a higher degree in the lower performance level achieved in the third test case. This test case involves eighty two test patterns in a somewhat wider range of figure distortion and aspect. The highest level of performance in this case stands at about 91 percent. These results are shown in the graph in Figure 13- 5.

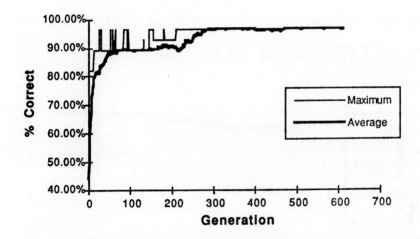

Figure 13-5. Performance on 28 test patterns.

		Circle	Square	Rectangle	Rhombus
1	Circle	0.789	0.751	0.717	0.512
2		0.797	0.748	0.694	0.503
3		0.801	0.747	0.673	0.507
4		0.779	0.679	0.582	0.418
5		0.806	0.734	0.620	0.501
6		0.807	0.731	0.599	0.503
7		0.810	0.729	0.586	0.504
8	Square	0.755	0.773	0.747	0.762
9		0.713	0.783	0.756	0.763
10		0.666	0.788	0.760	0.763
11		0.618	0.791	0.763	0.764
12		0.568	0.793	0.764	0.764
13		0.527	0.795	0.771	0.759
14		0.488	0.797	0.777	0.752

15	Rectangle	0.422	0.670	0.715	0.686
16		0.466	0.695	0.730	0.695
17		0.362	0.582	0.657	0.626
18		0.389	0.613	0.680	0.654
19		0.402	0.644	0.697	0.666
20		0.363	0.576	0.679	0.609
21		0.384	0.603	0.682	0.638
22	Rhombus	0.720	0.738	0.733	0.753
23		0.667	0.752	0.745	0.760
24		0.597	0.758	0.745	0.764
25		0.534	0.761	0.742	0.761
26		0.503	0.761	0.733	0.763
27		0.481	0.761	0.734	0.762
28		0.557	0.667	0.702	0.680

Table 13-1. Computed pattern membership grades for 28 test patterns.

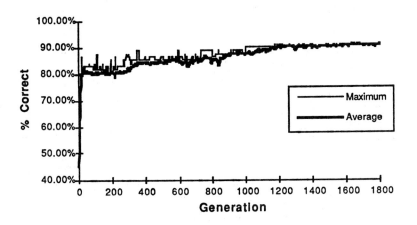

Figure 13-6. Performance on 82 test patterns.

13.5 CONCLUSION AND FUTURE DEVELOPMENT

Due to the fact that this recognition system was only recently made operational, initial results are still under analysis and hence, any conclusions drawn at this time are largely speculative. From the initial results we can however derive further procedures to both test and improve system performance.

The results of the current form of the system do however show considerable progress toward our original goals. Indeed, the system has demonstrated the fact that fuzzy sets can be used in pattern recognition to classify shapes using relatively simple and sparse pattern information. Furthermore, the fact that the initial configuration of the system has achieved above a 90 percent correct identification rate for a training cycle containing over eighty test patterns indicates that an effective and efficient recognition system can result from a merger of fuzzy logic and genetic algorithms.

Future work on this system concept includes generalization of the characteristic membership grades for the statistical components of the fuzzy measures. This will allow the genetic algorithm to choose the appropriate level of mean and standard deviation instead of comparing the pattern statistics with some expected ideal value. A second possibility for improvement involves a scheme to amplify in a nonlinear fashion, the current population of the coefficient vectors thereby improving population variation with more distinctive extreme points. Finally, we wish to investigate the possibility of deriving a special fitness criteria which would choose coefficient vectors that maximize relevant characteristics while minimizing those that are irrelevant. In other words, we want the pattern membership function to be high for correct figures and low for all others.

As previously mentioned, we have made considerable progress toward our overall project objectives. It is our hope that this beginning can be extended toward the ultimate goal of developing a comprehensive pattern recognition and analysis system for use in a broad range of applications such as computer vision and signal analysis.

REFERENCES

[1] Holland, J. H. (1962) Information Processing in Adaptive Systems. Information Processing in the Nervous System, *Proceedings of the International Union of Physiological Sciences*, 3, 330-339.

Holland, J. H. (1962) Outline for a logical theory of adaptive systems. *Journal of the Association for Computing Machinery*, 3, 297-314.

Holland, J. H. (1973) Genetic algorithm and the optimal allocations of trials. *SIAM Journal of Computing*, 2(2), 88-105.

Holland, J. H. (1981) Genetic algorithms and adaptation (Technical Report No. 34). Ann Arbor: University of Michigan, Department of Computer and Communications Sciences.

[2] Goldberg, D. E. (1989). *Genetic Algorithms in Search Optimization and Machine Learning*. Addison-Wesley.

14

FUZZY CONTROL OF ROBOTIC MANIPULATOR

*Kishan Kumar Kumbla, John Moya
and Ronald Baird
University of New Mexico*

14.1 INTRODUCTION

A robot is a reprogrammable multifunctional mechanical manipulator designed to move materials, parts, tools, or special devices through planned trajectories for the performance of a variety of tasks. It is a computer controlled manipulator consisting of several relatively rigid links connected in series by revolute, spherical, or translational joints. One of these links is typically attached to a *supporting base* while another link has an end that is free and equipped with a tool known as the *end-effector* for manipulating objects or performing assembly tasks. Mechanically, the robot is composed of an arm, wrist subassembly and a tool. It is designed to reach a workpiece located within some distance or workspace determined by the maximum and minimum elongations of the arm.

The dynamic equations of the robot are a set of highly nonlinear coupled differential equations containing a varying inertia term, a centrifugal and Coriolis term, a frictional term, and a gravity term. Movement of the end-effector in a particular trajectory at a particular velocity requires a complex set of torque functions to be applied by the joint actuators of the robot. The exact form of the required functions of actuator torque depend on the spatial and temporal attributes of the path taken by the end-effector as well as the mass properties of the links and payload, friction in the joints, etc.

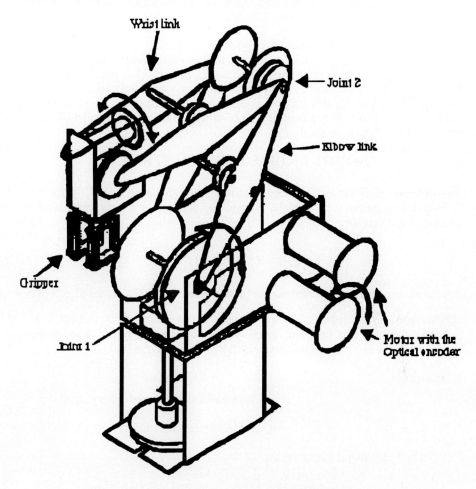

Figure 14-1. Schematic of a Rhino Robot.

The nonlinear dynamics governing robot motion presents a challenging control problem. A traditional linear controller cannot effectively control the motion of the robot. A controller based upon the theory of nonlinear control is better suited for the problems of robot manipulation. Unfortunately, nonlinear differential equations are plagued by substantial requirements for computation and have an incomplete theory of solution. Thus, most approaches to robot controller design have suffered due to the complications of nonlinear effects.

Because of these complications, fuzzy logic offers a very promising approach to robot controller design. Fuzzy logic offers design rules that are relatively easy to use in a wide range of applications, including nonlinear robotic equations. Fuzzy logic also allows for design in cases where models are incomplete, unlike most design techniques. In addition, microprocessor-based fuzzy controllers have performed with data streams of eight bits or less to allow for a simple design.

Section 14-2 describes the dynamic model of a robotic manipulator. A simulation of the fuzzy controller for a 2-link robotic manipulator is discussed. The corresponding simulation program is listed in Appendices 14-A and 14-B. Section 14-3 describes an implementation of fuzzy controller to control 3-links of a Rhino robot. The hardware developed for this purpose is described in detail. Section 14-4 describes the software developed in implementing this controller. Sections 14-5 and 14-6 discuss the results and conclusion, respectively.

14.2 ROBOT MODELS AND CONTROL

The dynamic equation of a robotic manipulator can be described by the nonlinear differential equation (1). For the purpose of simulation a two link planar robotic manipulator is considered [3].

$$M(\theta)\ddot{\theta} + C(\theta,\dot{\theta}) + G(\theta) = \tau \tag{1}$$

Where M is a two dimensional matrix of inertia terms, C is a vector of centrifugal and coriolis terms, G is a vector of gravity terms and τ is a vector of joint torques. $\theta, \dot{\theta}$ and $\ddot{\theta}$ are the joint angular position, velocity and acceleration terms. The mathematical expression for a two link robot is given by the following terms [5].

Inertia matrix:

$$M(\theta) = \begin{bmatrix} a_1 + a_2 \cos\theta_2 & a_3 + 0.5\, a_2 \cos\theta_2 \\ a_3 + 0.5\, a_2 \cos\theta_2 & a_3 \end{bmatrix} \tag{2}$$

Coriolis and centrifugal torque vector:

$$C(\theta,\dot{\theta}) = \begin{bmatrix} (a_2 \sin\theta_2)(\dot{\theta}_1 \dot{\theta}_2 + 0.5\dot{\theta}_2^2) \\ (a_2 \sin\theta_2) 0.5 \dot{\theta}_1^2 \end{bmatrix} \tag{3}$$

Gravity and loading vector:

$$G(\theta) = \begin{bmatrix} (a_4 \cos\theta_1) + a_5 \cos(\theta_1 + \theta_2) \\ a_5 \cos(\theta_1 + \theta_2) \end{bmatrix} \tag{4}$$

Chap. 14   Fuzzy Control of Robotic Manipulator   295

The parameters a_1, a_2, a_3, a_4 and a_5 account for the inertia and gravity that influences the motion nonlinear. Solutions to these nonlinear differential equations typically require several specialized techniques like Lyapunov theory, each one addressing a part of the solution, but none totally complete.

Fuzzy Logic Control

Fuzzy logic provides a means to deal with nonlinear functions. A fuzzy controller was designed to simulate the performance of the model of two-link Robotic manipulator. The membership functions were developed for the effect of position errors and velocity parameters for the two links of the robot. A membership function for the output of the controller i.e., the joint torque was also defined. This was developed using the Togai InfraLogic software [1]. The membership function of the fuzzy controller was first approximated by studying the response of a traditional PD controller and then finally tuned to achieve the best response by trial and error method. A set of 30 rules was developed. This is shown in Appendix 14-A. Thus, the membership function incorporate the characteristics and dynamics of the manipulator.

Implementation

The dynamics of the system is simulated using a software developed in C language. The program is listed in Appendix 14-B. The main function first defines the various global and local variables; initial and final time of the simulation. The *des_traj()* subroutine defines the desired trajectory (step input) and the *controller()* subroutine gives the output of the fuzzy controller. It then calls the subroutine *robo_state()* which defines the dynamic of the two link manipulator. The dynamics of the robot is solved by the subroutine *rk4()*, which is basically a 4th order Runge Kutta solution method. The *out_file()* converts the output in a suitable form that can be used to generate a *matlab* compatible file by a *print_matlab()* subroutine. This forms the driver program. The *.til* file which defines the fuzzy controller in Togai InfraLogic software is first converted into C code and merged with the driver program and then compiled and linked to form the executable code. The output data file generated by the simulation is used to plot the trajectory using *matlab* software.

Simulation Results

Figure 14-2 shows a comparison of fuzzy controller vs the PD controller. A response to a step input is plotted against time. Figure 14-2a shows the fuzzy controller response and Figure 14-2b shows the PD controller response. Simulation shows that fuzzy controller has a rise time of 0.3 seconds, whereas the PD controller has a rise time of 0.4 sec. Fuzzy controller is able to improve the response by reducing the overshoot and at the same time decreasing the rise time. Note that various sets of gains were tried in the case of PD controller to obtain an optimal performance. It has been seen that if the gains of the PD controller is increased to improve the rise time, there is an inherent increase in the rise time. Also, one set of gains of the PD controller do not provide the same performance in all the operating ranges.

Fuzzy logic attempts to bypass the difficulties that accompany the solutions to nonlinear differential equations. Fuzzy logic allows us to set-up a controller that is not entirely based on a complete description of the robot.

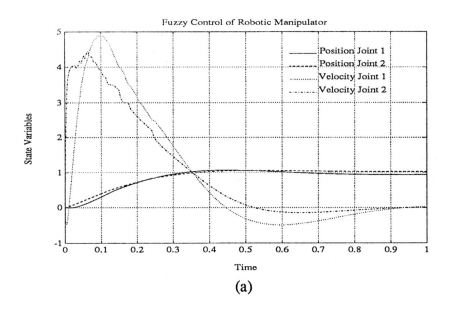

(a)

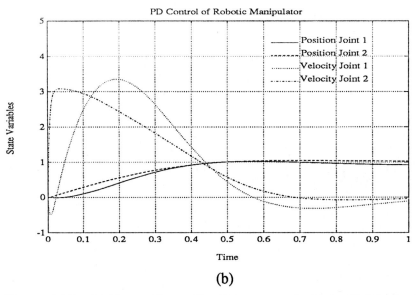

(b)

Figure 14-2. (a) Fuzzy Controller Response of a Robotic Manipulator. (b) PD Controller Response of a Robotic Manipulator.

14.3 BASELINE HARDWARE DESIGN

Figure 14-3 provides an overall block diagram of the Rhino robot and control hardware. The robot has five links and a gripper. The three links used for controlled movement included the shoulder, elbow, and wrist. Each link was moved with a DC servo motor rotating about an axis at each joint. The computer system used was a Packard Bell 386PC. The remaining hardware consisted of elements for power control (driver), data acquisition, and communications.

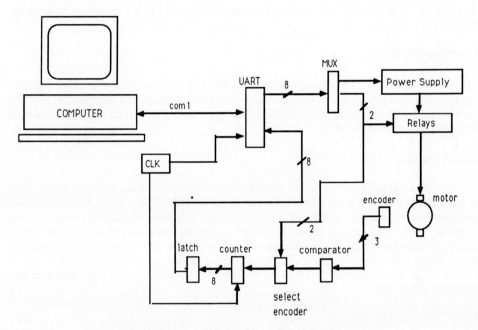

Figure 14-3. Schematic of the Hardware.

Power Control

The driver for the DC motors consisted of a Fluke 4265A programmable power supply, control circuits and relays. Two eight-bit control words were used to control the driver. The control words were originally sent by the computer through the RS-232C COM1 port in serial format to the external universal asynchronous receiver/transmitter (UART). The UART then delivered the control words in parallel to the programmable power supply and MUX. The first eight-bit control word was used to program the power supply voltage level, thus allowing the supply to deliver voltages ranging from —16 to +16 volts in 0.25 volt increments. The second eight-bit control word controlled the time that the power supply switched output settings and signaled the MUX to activate one of the relays. The relays were required to switched power to only one motor at a time. Thus, at any point in time, only one of the links could be activated.

Data Acquisition

The data acquisition hardware provided the position feedback required by the software controller. Optical encoders were used to detect rotational movement of a motor and send six analog pulses for one complete revolution. Each pulse corresponded to 0.12 degrees of movement. The analog pulses then passed through a Schmitt-trigger comparator to give a clean pulse for digital circuit interface. The pulses incremented the counter and thus provided a measurement of total displacement. At every ten milliseconds a crystal clock triggered the sampling of the counter. The count information was latched, loaded into the UART as parallel data, and then transmitted serially to the computer through the RS-232C COM1 port.

Communications

The communications hardware provided the necessary elements for transporting data between the computer and external hardware. The primary hardware components used for communications included the COM1 serial port and the Universal Asynchronous Receiver Transmitter (UART).

COM1 port

The COM1 port is a serial, asynchronous communications device that comes as a standard accessory with IBM PCs and most compatibles. The COM1 port was supported by a software driver provided with the MS-DOS operating system. Data entering the COM1 port is made available to the software hosted on the computer. Likewise, data generated by the software may be sent to devices external to the computer through the COM1 port interface. The data entering or leaving the port had to conform to the RS-232C standard. The port was set to operate at 9600 baud with eight data bits, two stop bits and no parity.

UART Implementation

The UART shown in Figure 14-4 was a 40 pin TR1602 by Western Digital Corporation. Four of the pins were available for RS-232C option settings. The NO PARITY pin (pin 35) was asserted high to eliminate the presence of the parity bit in the serial data. The NUMBER OF STOP BITS pin (pin 36) was asserted high to designate two stop bits for the serial data. The pins for NUMBER OF DATA BITS (pin 37 and pin 38) were both asserted high to set the number of data bits to eight. The other possible choices for NUMBER OF DATA BITS were five, six, and seven data bits. Eight more pins on the UART were dedicated for TTL compatible, eight-bit parallel data interface [4].

UART Transmitter Configuration

The count information arrived at the UART as a parallel eight-bit data word awaiting to be transmitted serially to the COM1 port. When the DATA STROBE pin

Chap. 14   Fuzzy Control of Robotic Manipulator   299

(pin 23) was asserted low, it signaled the UART to load the eight-bit data word into its transmitter buffer register. The TRANSMITTER BUFFER EMPTY pin (pin 22) provided a signal to the COM1 port indicating that the transmitter buffer register was empty (logic one). The END OF CHARACTER pin (pin 24) provided a signal to the COM1 port indicating when transmission was stopped.

During operation, the UART first transmitted TTL compatible serial data through the SERIAL OUTPUT pin (pin 25) to the MC1488 IC. The MC1488 then converted the TTL signal to RS-232C format that was compatible with the COM1 port.

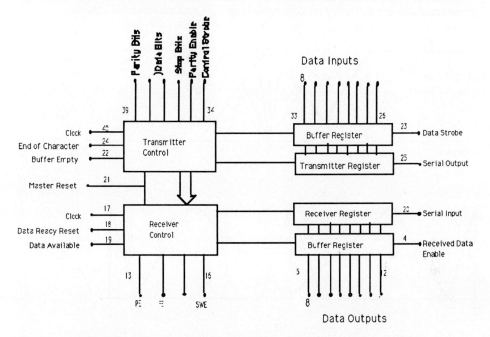

Figure 14-4.   Functional Block Diagram of the UART.

UART Receiver Configuration

Serial data originating from the COM1 port conformed to the RS-232C standard. The serial data then passed through the MC1489 where it was converted to a signal that was TTL compatible for interface with the UART. The TTL compatible serial data then arrived at the SERIAL INPUT pin (pin 20) of the UART. The UART converted the serial data into parallel eight-bit data words capable of interfacing with the driver for the motors.

Bit Rate Generator

The UART required a stable clock signal to be applied to the CLOCK pin (pin 40) in order to transmit and receive serial data precisely at 9600 baud. Furthermore, the clock signal was required to operate at 16 times the frequency of the baud rate. The clock

300 Chap. 14   Fuzzy Control of Robotic Manipulator

signal used was generated with a Motorola MC14411 bit rate generator IC, which essentially is a frequency divider. A five volt source connected to the bit rate generator enabled the clock to pulse at a rate of 1.832 MHz using a crystal oscillator. Different frequencies were available from the bit rate generator IC for different parts of the system.

14.4 SOFTWARE IMPLEMENTATION OF THE FUZZY CONTROLLER

The controller software was divided into two primary modules written in the C language. The two modules are the fuzzy control module and the communications module.

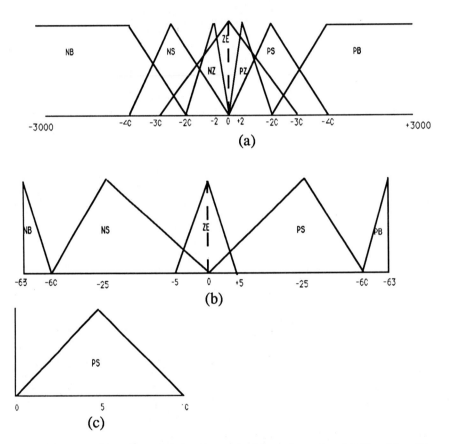

Figure 14-5. Fuzzy Membership Functions for (a) Position Error, (b) Output Voltage, and (c) Velocity.

Fuzzy Control Module

The first step in fuzzy controller design was the selection of input and output variables. The designated input variables were ERROR (robot link position error) and VELOCITY

Chap. 14 Fuzzy Control of Robotic Manipulator 301

(velocity of the robot link). The output variable was chosen to be VOLTAGE (drive voltage output by the power supply). Each variable was accompanied with a set of membership functions. The membership functions are shown in Figure 14-5.

During actual operation, the computer would read in the robot position data and then compute position error and velocity. The fuzzy controller fuzzified the input quantities through algorithms that operated on the input data as specified by the membership functions. Next, the fuzzified input quantities passed through a series of IF-THEN decision rules that formed the main body of the fuzzy controller. Thus, the fuzzy controller routinely assessed the current state of the robot and determined which control action was most appropriate. Defuzzification was applied using the voltage output variable and a control action was selected. The control action was then passed through the interface to the external power supply driver.

Function	**Code**
Shoulder Relay	MSB 1010 LSB 0000
Elbow Relay	MSB 1011 LSB 0000
Wrist Relay	MSB 1010 LSB 1000
Positive Sign	MSB 1001 LSB 1000
Negative Sign	MSB 1001 LSB 0000

Table 14-1. Control words for the Driver.

The output control signal from the computer is made up of two eight-bit control words that occupied two control cycles for each action completed. The first control word was used to either select one of the three relays or to change the voltage polarity output of the programmable power supply. This action was identified by bit seven, the most significant bit (MSB), and bit six. When bit seven was set to one and bit six set to zero, then either one of the relays was selected or the voltage polarity output of the power supply was set to the desired output. Bits five, four and three were used to designate which relay to be select or to set the output voltage polarity. When bit seven was reset to zero, the control circuit then activated one of the motors of the robot. Bit seven was used to strobe the programmable power supply and thus activate the programmed voltage output. The codes for the different actions are shown in Table 14-1.

The second control word was identified with both bit seven and bit six set to one. The desired voltage level output of the programmable power supply was selected by bits zero through five. Bit seven was then set to zero in the next eight-bit control word to strobe the power supply and thus activate the newly selected power supply output voltage. Correct power supply operation required the use of two's-complement arithmetic with inverted logic.

The design of the fuzzy controller algorithm was accomplished with the Togai InfraLogic's Fuzzy-C Compiler [1]. Input to the Togai Fuzzy-C Compiler consisted of a fuzzy source code with membership functions and a set of IF-THEN decision rules. This is listed in Appendix 14-C. The Fuzzy-C Compiler converted the fuzzy source code into standard C source code which then passed through a C compiler to produce an executable code of the fuzzy controller. Trial experiments of the executable code were conducted to calibrate the membership functions until robot trajectory overshoot was suppressed and rise time was kept to a minimum.

Communications Module

A communications routine sampled the count information in the receive register every ten milliseconds. The fuzzy controller operated on this data and then outputted the results to a transmit register which was used to drive the DC motors. The program is listed in Appendix 14-D.

14.5   TESTING AND RESULTS

Two tests were conducted to compare controller performance on robot arm trajectory. In both tests, the gain of the PD controller was set to provide the best performance in all ranges of operation. In the first test, each controller was set to traverse a trajectory of 200 counts in both positive (clockwise) and negative (counterclockwise) directions (Note: One count is equal to 0.12 degrees of rotation of the link). The test results for the trajectory of the shoulder, elbow and wrist are shown in Figure 14-6. The results show that the fuzzy controller is able to move the robot arm smoothly to the desired position without overshoot. The PD controller, however, generated overshoot in the trajectory. It is interesting to note that when the robot link approached its final destination, its velocity remained high (steep trajectory curve) with PD control, but began to decrease much earlier with fuzzy control.

In the second test, the controllers were compared under various ranges of operation. The PD controller with suitable gain improved trajectory response by reducing the overshoot to less than four counts. However, the PD controller was unable to move the motor when the links were initially only one or two counts away from the final destination. The primary cause for this effect was the nonlinear friction in the robot joints. One could overcome this resistance with an increase in controller gain, but then the existing trajectory overshoot would be intensified.

The fuzzy controller was able to overcome the problems encountered by traditional PD control and perform better in all conditions tested. Results showed that the maximum overshoot with fuzzy control was held to a count of one. The fuzzy controller also was able to actuate the motor and travel a short distance of one count, unlike the PD controller.

The fuzzy controller exhibited robustness in performance against non-ideal effects like robot inertia, Coriolis effect and gravity. These effects influence the velocity, position, and acceleration of the robot links, and thus negatively impact controller performance. However, with fuzzy control, only one set of fuzzy membership functions was required to guard against these non-ideal effects and operating all links of the robot. This result is not shared by the conventional PD controller which required different gains to accommodate for the differences between each link.

14.6 CONCLUSIONS

The fuzzy controller was able to suppress the robot trajectory overshoot and perform better than traditional PD control under all conditions tested. Furthermore, one set of fuzzy membership functions was sufficient to accommodate the variations that occurred in operating and controlling the different links of the robot. This implies that fuzzy control is robust and able to adapt to many unforeseen elements inevitable in any practical implementation.

REFERENCES

[1] Togai InfraLogic Inc. *Fuzzy-C Development System User's Manual*, Irvine, CA, 1991.
[2] Rhino, Inc. *Rhino Robot XR Series Owner's Manual*, Champaign, IL, 1988.
[3] Craig, John J. *Introduction to Robotics*. Addison-Wesley, Reading, Mass., 1989.
[4] Larsen, D. G., Rony, P. R., Titus, J. A., and Titus, C. A. *Interfacing and Scientific Data Communications Experiments*. Howard W. Sams & Co., Inc., Indianapolis, Indiana, 1979.
[5] Jet Propulsion Laboratory New Technology Report. Methods and Apparatus for Adaptive Force and Position Control of Manipulator. JPL & NASA Case No. NPO-17127.
[6] Jamshidi, M., Barak, D., Vadiee, N., and Baugh, S. A Simulation Environment for Adaptive Fuzzy Control Systems. CAD Laboratory for Intelligent and Robotic Systems, Department of Electrical and Computer Engineering, University of New Mexico, Albuquerque, NM.

Appendix 14-A

```
/* Two-link Robot Control using Fuzzy Logic.
   Membership functions.                       */

PROJECT robot_control

VAR e_theta1
       TYPE float
       MIN -3.15
       MAX 3.15
   MEMBER ZE
             POINTS -0.5 0 0 1 0.5 0
       END
       MEMBER PS
             POINTS 0 0 0.5 1 1 0
       END
       MEMBER NS
             POINTS 0.0 0 -0.5 1 -1 0
       END
       MEMBER PM
             POINTS 0.5 0 1 1 3.15 1
       END
       MEMBER NM
             POINTS -0.5 0 -1 1 -3.15 1
       END
END

VAR e_theta2
       TYPE float
       MIN -3.15
       MAX 3.15

   MEMBER ZE
             POINTS -0.5 0 0 1 0.5 0
       END
       MEMBER PS
             POINTS 0 0 0.5 1 1 0
       END
       MEMBER NS
             POINTS 0.0 0 -0.5 1 -1 0
       END
       MEMBER PM
             POINTS 0.5 0 1 1 3.15 1
       END
       MEMBER NM
             POINTS -0.5 0 -1 1 -3.15 1
       END
```

Chap. 14 Fuzzy Control of Robotic Manipulator

```
END
VAR dtheta1
        TYPE float
        MIN -5.0
        MAX 5.0
        MEMBER ZE
                POINTS -1 0 0 1 1 0
        END
        MEMBER PS
                POINTS 0 0 1 1 5 1
        END
        MEMBER NS
                POINTS -5 1 -1 1 0 0
        END
END
VAR dtheta2
        TYPE float
        MIN -5.0
        MAX 5.0
        MEMBER ZE
                POINTS -1 0 0 1 1 0
        END
        MEMBER PS
                POINTS 0 0 1 1 5 1
        END
        MEMBER NS
                POINTS -5 1 -1 1 0 0
        END
END
VAR torque1
        TYPE float
        MIN -2000
        MAX 2000
        MEMBER ZE
                POINTS -200 0 0 1 200 0
        END
        MEMBER PS
                POINTS 0 0 200 1 400 0
        END
        MEMBER NS
                POINTS -400 0 -200 1 0 0
        END
        MEMBER PM
                POINTS 200 0 400 1 600 0
        END
        MEMBER NM
                POINTS -600 0 -400 1 -200 0
        END
        MEMBER PL
```

```
                POINTS 400 0 600 1 2000 1
        END
        MEMBER NL
                POINTS -2000 1 -600 1 -400 0
        END
END
VAR torque2
        TYPE float
        MIN -2000
        MAX 2000
        MEMBER ZE
                POINTS -200 0 0 1 200 0
        END
        MEMBER PS
                POINTS 0 0 200 1 400 0
        END
        MEMBER NS
                POINTS -400 0 -200 1 0 0
        END
        MEMBER PM
                POINTS 200 0 400 1 600 0
        END
        MEMBER NM
                POINTS -600 0 -400 1 -200 0
        END
        MEMBER PL
                POINTS 400 0 600 1 2000 1
        END
        MEMBER NL
                POINTS -2000 1 -600 1 -400 0
        END
END
FUZZY control_rules
        RULE rule1 IF e_theta1 IS NM AND dtheta1 IS PS THEN
                torque1 IS NL END
        RULE rule2
                IF e_theta1 IS NM AND dtheta1 IS ZE THEN
                        torque1 IS NM
        END
        RULE rule3
                IF e_theta1 IS NM AND dtheta1 IS NS THEN
                        torque1 IS NS
        END
        RULE rule4
                IF e_theta1 IS NS AND dtheta1 IS PS THEN
                        torque1 IS NM
        END
        RULE rule5
                IF e_theta1 IS NS AND dtheta1 IS ZE THEN
```

```
                torque1 IS NS
END
RULE rule6
        IF e_theta1 IS NS AND dtheta1 IS NS THEN
                torque1 IS ZE
END
RULE rule7
        IF e_theta1 IS ZE AND dtheta1 IS PS THEN
                torque1 IS NS
END
RULE rule8
        IF e_theta1 IS ZE AND dtheta1 IS ZE THEN
                torque1 IS ZE
END
RULE rule9
        IF e_theta1 IS ZE AND dtheta1 IS NS THEN
                torque1 IS PS
END
RULE rule10
        IF e_theta1 IS PS AND dtheta1 IS PS THEN
                torque1 IS ZE
END
RULE rule11
        IF e_theta1 IS PS AND dtheta1 IS ZE THEN
                torque1 IS PS
END
RULE rule12
        IF e_theta1 IS PS AND dtheta1 IS NS THEN
                torque1 IS PM
END
RULE rule13
        IF e_theta1 IS PM AND dtheta1 IS PS THEN
                torque1 IS PS
END
RULE rule14
        IF e_theta1 IS PM AND dtheta1 IS ZE THEN
                torque1 IS PM
END
RULE rule15
        IF e_theta1 IS PM AND dtheta1 IS NS THEN
                torque1 IS PL
END
RULE rule16
        IF e_theta2 IS NM AND dtheta2 IS PS THEN
                torque2 IS NL
END
RULE rule17
        IF e_theta2 IS NM AND dtheta2 IS ZE THEN
                torque2 IS NM
```

END
RULE rule18
    IF e_theta2 IS NM AND dtheta2 IS NS THEN
        torque2 IS NS
END
RULE rule19
    IF e_theta2 IS NS AND dtheta2 IS PS THEN
        torque2 IS NM
END
RULE rule20
    IF e_theta2 IS NS AND dtheta2 IS ZE THEN
        torque2 IS NS
END

RULE rule21
    IF e_theta2 IS NS AND dtheta2 IS NS THEN
        torque2 IS ZE
END

RULE rule22
    IF e_theta2 IS ZE AND dtheta2 IS PS THEN
        torque2 IS NS
END
RULE rule23
    IF e_theta2 IS ZE AND dtheta2 IS ZE THEN
        torque2 IS ZE
END
RULE rule24
    IF e_theta2 IS ZE AND dtheta2 IS NS THEN
        torque2 IS PS
END
RULE rule25
    IF e_theta2 IS PS AND dtheta2 IS PS THEN
        torque2 IS ZE
END
RULE rule26
    IF e_theta2 IS PS AND dtheta2 IS ZE THEN
        torque2 IS PS
END
RULE rule27
    IF e_theta2 IS PS AND dtheta2 IS NS THEN
        torque2 IS PM
END
RULE rule28
    IF e_theta2 IS PM AND dtheta2 IS PS THEN
        torque2 IS PS
END
RULE rule29
    IF e_theta2 IS PM AND dtheta2 IS ZE THEN

                              torque2 IS PM
        END
        RULE rule30
                IF e_theta2 IS PM AND dtheta2 IS NS THEN
                        torque2 IS PL
END
END
/* The following CONNECT Objects specify that e_theta1, e_theta2, dtheta1
and dtheta2 are the inputs to the Control_rules knowledge base while
torque1 and torque2 are the outputs from the Control_rules. */

CONNECT
        FROM e_theta1
        TO control_rules
END
CONNECT
        FROM e_theta2
        TO control_rules
END
CONNECT
        FROM dtheta1
        TO control_rules
END
CONNECT
        FROM dtheta2
        TO control_rules
END
CONNECT
        FROM control_rules
        TO torque1
END
CONNECT
        FROM control_rules
        TO torque2
END
END

Appendix 14-B

/***/

/* CONTROL OF A TWO LINK ROBOT USING FUZZY LOGIC. */

/***/

```c
#include <stdio.h>
#include <math.h >
#include <stdlib.h>

/* define the global variables */
 float des_x[4];  /* Desired trajectory vector.*/

int nt=0;
 float big_time[1000];
 float big_state[1000][4];

/* The main program starts .*/
main()
{
int i;
int n=4;
float ini_time,fin_time;  /* initial time and final time. */
float time,h;      /* simulation time and integation step interval */
extern float des_x[];
float s_x[4],d_s_x[4],err[4],torque[2];/* state vector,next state,error */
float s_x1[4]; /* Basically the state vector. */
float x_ini[4]; /* Initial condition */

void robo_state();
void rk4();
void output_file();

printf(" Enter the initial time.\n");
scanf("%f", &ini_time);
printf(" Enter the final time. \n");
scanf("%f", &fin_time);

 /* Calculate the step size of integration. */

h=(fin_time-ini_time)/100.0;

 /* Set the initial states of the robot. Normally the zero vector.*/
         x_ini[0]=0.0;
         x_ini[1]=0.0;
```

Chap. 14 Fuzzy Control of Robotic Manipulator   311

```
        x_ini[2]=0.0;
        x_ini[3]=0.0;
        for(i=0;i<n;i++) s_x[i]=x_ini[i];
        time=ini_time;
        while( time < fin_time)
{
        des_traj(des_x);      /* Get the desired trajectory. */
        controller(s_x,des_x,torque,err); /* Get the controller output.*/
        /* The controller output is the output of the fuzzy controller.*/
        /*  ctrl.c is the c version of the controller file,       */
        robo_state(time,s_x,torque,d_s_x);/* The robot dynamics. */

        /* Solve the robot equation using Runge Kutta method. */
        rk4(n, time, h, s_x, d_s_x, s_x1,torque);
        for(i=0;i<n;i++) s_x[i]=s_x1[i] ;
        output_file(time,s_x);

        /* output every interval to create a big array of time and state.*/
        time=time+h; /* Increment the time. */
} /* end of the simulation time. */

printf(" Number of iterations := %d \n",nt);
print_matlab(); /* writes the output in a matlab compatible file, */
} /* end of the main program. */

/***************************************************************/

#define m1 15.91    /* define the robot constants, */
#define m2 11.36
#define len 0.432
#define grv 9.81
#include <math.h>

    /* Subroutine to define the robot dynamics. */
    void robo_state(tim1,s1_x,torque1,d1_s_x)

        float tim1,s1_x[],torque1[],d1_s_x[];
        {
          /* Local variables. */
        float a1,a2,a3,a4,a5;
        float m_11,m_12,m_21,m_22;
        float invm_11,invm_12,invm_21,invm_22;
        float co_1,co_2;
        float gr_1,gr_2;
        float deter;
        float t_fun1,t_fun2,fun1,fun2;

tim1=tim1; /* dummy statement */
```

```
/* Robot contants. */
a1=m2*len*len +len*len *(m1+m2);
a2=2.0*len*len*m2;
a3=len*len*m2;
a4=(m1+m2)*len*grv;
a5=m2*len*grv;

/* Define the robot equation parameter */
/* Inertia Matrix elements */
m_11=a1+a2*cos(s1_x[1]);
m_12=a3+0.5*a2*cos(s1_x[1]);
m_21=m_12;
m_22= a3;

/* Calculate the inverse of the inertial matrix */
deter=m_11*m_22 - m_12*m_21; /* determinent of inertia matrix */

invm_11=m_22/deter;
invm_12=-m_12/deter;
invm_21=-m_21/deter;
invm_22=m_11/deter;

/* Corilis matrix elements */
co_1=-(a2*sin(s1_x[1]))*(s1_x[2]*s1_x[3]+s1_x[2]*s1_x[2]*0.5);
co_2=(a2*sin(s1_x[1])*s1_x[2]*s1_x[2]*0.5);

/* Gravity elements */
gr_1=a4*cos(s1_x[0])+a5*cos(s1_x[0]+s1_x[1]);
gr_2=a5*cos(s1_x[0]+s1_x[1]);

t_fun1=torque1[0]-co_1-gr_1;
t_fun2=torque1[1]-co_2-gr_2;
fun1=invm_11*t_fun1 + invm_12*t_fun2;
fun2=invm_21*t_fun1 + invm_22*t_fun2;

/* The State Equations */
d1_s_x[0]=s1_x[2];
d1_s_x[1]=s1_x[3];
d1_s_x[2]=fun1;
d1_s_x[3]=fun2;

}

/*************************************************************/

/* Runge Kutta subroutine */
void rk4(no,x,T,y,dydx,yout,torque1)
        int no;
        float x,T,y[],dydx[],yout[],torque1[];
```

Chap. 14   Fuzzy Control of Robotic Manipulator   313

```
                /*  void (*derivs)();*/
                /* void (*derivs)(float,float *,float *,float *); */
                {
                int test;
                int i;
                float xh,hh,h6,dym[4],dyt[4],yt[4];

hh=T*0.5;
h6=T/6.0;
xh=x+hh;
for (i=0;i<no;i++) yt[i]=y[i]+hh*dydx[i];
robo_state(xh,yt,torque1,dyt);
for (i=0;i<no;i++) yt[i]=y[i]+hh*dyt[i];
robo_state(xh,yt,torque1,dym);
for (i=0;i<no;i++)
        {
        yt[i]=y[i]+T*dym[i];
        dym[i] += dyt[i];
        }
robo_state(x+T,yt,torque1,dyt);
for (i=0;i<no;i++)
        {
        yout[i]=y[i]+h6*(dydx[i]+dyt[i]+2.0*dym[i]);
        }
}
/***************************************************************/

  /* Desired trajectory generator. */

  des_traj()
  /*  float des_x1[];*/
  {
  des_x[0]=1.0;   /*step input. */
  des_x[1]=1.0;
  des_x[2]=0.0;
  des_x[3]=0.0;
  }

/***************************************************************/

/* The controller routine. */

controller(s1_x,des1_x,torque1,err1)
float s1_x[],des1_x[],torque1[],err1[];
{
            int i;
            float kp1;
            float kp2,kv1,kv2;
```

```
kp1=680.0;
kp2=550.0;
kv1=100.0;
kv2=90.0;

for(i=0; i<4; i++)
        err1[i]= des1_x[i] - s1_x[i];

torque1[0]=kp1 * err1[0]+kv1 * err1[2];
torque1[1]=kp2 * err1[1]+kv2 * err1[3];
}
```

/**/

/* Output routine. This open a file and puts the
   value of time and S_x[] in a Matlab readable format. */

```
void output_file(tim1,s1_x)
        float tim1, s1_x[];
        {
        extern int nt;
        extern float big_state[][],big_time[];
        int j;
big_time[nt]=tim1;
for(j=0;j<=4;j++)
big_state[nt][j]=s1_x[j];
        nt++;
out=fopen("output.m","a");
fprintf(out, "[ %f ] [ %f %f %f %f ]\n", tim1,s1_x[0],s1_x[1],
                                        s1_x[2],s1_x[3]);
 fclose(out);
}
```

/**/

```
print_matlab()
{
        FILE *out;
        extern int nt;
        extern float big_time[], big_state[][];
        int ct;

out=fopen("out_kis.m","a");
fprintf(out, "\n %% This is an output of the robot dynamic simulation. \n");
fprintf(out,"Time= [ ");
for(ct=0;ct<nt;ct++)
        fprintf(out,"%f;\n ",big_time[ct]);
        fprintf(out," ]\n\n");
```

```
        fprintf(out," State Vector = [ \n");
        for(ct=0;ct<nt;ct++)
        fprintf(out,"%f %f %f %f ;\n",big_state[ct][0],big_state[ct][1],
                                     big_state[ct][2],big_state[ct][3]);

        fprintf(out," ]\n\n");
}

/**************************************************************/
```

Appendix 14-C

```
PROJECT controller
VAR vel
TYPE signed word
MIN 0
MAX 10
            MEMBER PS
                        POINTS 0 1 5 1 10 0
            END
END
VAR error
TYPE signed word
MIN -3000
MAX  3000
            MEMBER PZ
                        POINTS 0 0 2 1 20 0
            END
            MEMBER NZ
                        POINTS -20 0 -5 1 0 0
            END

            MEMBER ZE
                        POINTS -30 0 0 1 30 0
            END
            MEMBER PS
                        POINTS 0 0 20 1 40 0
            END
            MEMBER NS
                        POINTS -40 0 -20 1 0 0
            END
            MEMBER PB
                        POINTS 20 0 40 1 3000 1
            END
            MEMBER NB
                        POINTS -3000 1 -40 1 -20 0
            END
END

VAR voltage
TYPE signed word
MIN -63
MAX  63
            MEMBER ZE
                        POINTS -5 1 0 1 5 1
            END
            MEMBER PS
                        POINTS 0 0 25 1 60 0
```

Chap. 14 Fuzzy Control of Robotic Manipulator

```
                END
                MEMBER NS
                        POINTS -60 0 -25 1 0 0
                END
                MEMBER PB
                        POINTS  60 0 63 1
                END
                MEMBER NB
                        POINTS -63 1 -60 0
                END
END
FUZZY controller_rules
        RULE rule1
                IF error IS PB THEN voltage IS NB
        END
        RULE rule2
                IF error IS PS THEN voltage IS NS
        END
        RULE rule3
                IF error IS ZE THEN voltage IS ZE
        END
        RULE rule4
                IF error IS PZ AND vel IS PS THEN voltage IS NS
        END
        RULE rule5
                IF error IS NZ AND vel IS PS THEN voltage IS PS
        END
        RULE rule6
                IF error IS NS THEN voltage IS PS
        END
        RULE rule7
                IF error IS NB THEN voltage IS PB
        END
END

        CONNECT
        FROM error
        TO controller_rules
        END
        CONNECT
        FROM vel
                TO controller_rules
        END
        CONNECT
        FROM controller_rules
        TO voltage
        END
END
```

Appendix 14-D

```
/***************************************************
 *
 *   FUZZY CONTROL OF 3 LINKS OF A RHINO ROBOT.
 *
 * **************************************************/

#include <conio.h>
#include <stdio.h>
#include <stdlib.h>
#include <dos.h>
#include <ctype.h>

#define srl_data  0x03f8
#define srl_stat  0x03fd
#define NOP       0xff
#define LIMIT     9999
#define ON 1
#define OFF 0
#define POS 1
#define NEG 0
#define MAX 63
#define MIN -63
#define N 999
/* define global variables */
int sh_p[N],el_p[N],wr_p[N];
int i=0;
int j=0;
int k=0;
int sh_pos,el_pos,wr_pos;
int des_sh_pos,des_el_pos,des_wr_pos;
int shoulder= OFF;
int elbow   = OFF;
int wrist   = OFF;
int sign    = POS;
int data_pre;

    /* The main Program. */

main()
{

printf("Is the Rhino in Hard Home Position? (Y/N) :");
if (toupper(getch())=='Y')
            ini_zero();
else {
         printf("Initialize Rhino to last run position? (Y/N) :\n");
```

Chap. 14   Fuzzy Control of Robotic Manipulator   319

printf("If NO the Rhino is initialized to zero.");

if(toupper(getch())=='Y') upd_last();

else ini_zero();
}
printf("\nENTER DESIRED POSITION (shoulder elbow wrist :\n");
scanf("%d %d %d",&des_sh_pos,&des_el_pos,&des_wr_pos);
printf("Do you want to move the robot to the desired position?\n");

if (toupper(getch())!='Y') exit1();

set_data(); /* set the data format. */
set_baud(); /* set the baud rate 9600. */

move();

/* move the links to the desired position,. */

exit1();

/* store the current position in the file before exiting . */

} /* end of main */

/**/

move()   /* move the robot to the desired position. */
{
        data_pre=inp(srl_data);
        move_shl();   /* move shoulder link. */ data_pre=inp(srl_data);
        move_elb();   /* move elbow link. */ data_pre=inp(srl_data);
        move_wrt();   /* move the wrist link. */
} /* end of move. */

/**/
    move_shl()   /* move shoulder link. */
        {
            void controller();
            int shl_v=0;
                float kp=8.0;
                float kd=2.0;
            int cont_out;
                int shl_errp1=0;
                int shl_errp2=0;
                int volt;
            shoulder= ON;

```c
        s_relay1();/* turn the relay connecting shoulder motor p.s on */

        do
        {
           sample(shl_v); /* sample the encoder output. */

           shl_errp1=err_sh();  /* calculate the error */

    /* Call the fuzzy controller . */

    controller(shl_errp1,shl_v,&cont_out);

    /* cont_out= -kp*shl_errp1-kd*shl_v;  */

                /* set the MAX and MIN values of the controller.*/
                if (cont_out> MAX)
                    cont_out=MAX;
                else if (cont_out < MIN)
                        cont_out=MIN;
                if (cont_out < 0)
                {
                sign=NEG;
                volt=-cont_out;
                }

                else
                {
                sign=POS;
                volt=cont_out;
                }

           output(cont_out); /* output to set the voltage of the p.s. */
           shl_errp2=shl_errp1;
           sample_delay(); /* sampling time adjustment. */
        }while(volt >= 3);
        s_0volts();    /* turn of the p.s. */
        shoulder=OFF;
        }
           /* end of shoulder movement. */
/*****************************************************/

        move_elb()   /* elbow move. */
        {

           int elb_v=0;
                float kpe=2.8;
                float kde=1.0;
```

Chap. 14 Fuzzy Control of Robotic Manipulator

```
        int cont_out;
        int elb_errp;
              int volt;
        elbow= ON;
          s_relay2();
        do
        {
        sample(elb_v);
           elb_errp=err_el();
/* Call the fuzzy controller . */
controller(elb_errp,elb_v,&cont_out);
/*cont_out= -kpe*elb_errp-kde*elb_v;*/
                                   /* PD controller */
        if (cont_out> MAX)
                 cont_out=MAX;
            else if (cont_out < MIN)
                      cont_out=MIN;
            if (cont_out < 0)
            {
            sign=NEG;
            volt=-cont_out;
            }
            else
            {
            sign=POS;
            volt=cont_out;
            }
         output(cont_out);
         sample_delay();
         } while(volt >=3);
         s_0volts();
         elbow=OFF;
         }    /* end of elbow move. */
```

/**/

```
     move_wrt()    /* wrist move */
     {

       int wrt_v=0;
             float kpw=2.8;
             float kdw=1.0;
       int wrt_errp;
                      unsigned int volt;
       int cont_out;
       wrist= ON;
         s_relay3();
         do
```

```
        {
    sample(wrt_v);
            wrt_errp=err_wr();
/* Call the fuzzy controller . */
    controller(wrt_errp,wrt_v,&cont_out);
    /* cont_out= -kpw*wrt_errp-kdw*wrt_v;*/

    if (cont_out> MAX)
            cont_out=MAX;
        else if (cont_out < MIN)
                cont_out=MIN;
        if (cont_out < 0)
        {
        sign=NEG;
        volt=-cont_out;
        }
            else
      {
        sign=POS;
        volt=cont_out;
        }
    output(cont_out);
    sample_delay();
    } while (volt >=3);
    s_0volts();
    wrist=OFF;
    }
```

/**/

/* sample the receiver buffer and read the counter output. */

```
            sample(increment)
            int increment;

            {
                int data_now;
                data_now=inp(srl_data);
                if (data_now < data_pre)
                        increment= 256 - data_pre + data_now;
                else increment= data_now - data_pre;

                if (shoulder==ON )
                    {
                            if (sign==NEG)      /* record the movement. */
                        sh_pos+=increment;
                            else sh_pos-=increment;
                        sh_p[i]=sh_pos;
```

Chap. 14 Fuzzy Control of Robotic Manipulator

```
                    i++;
                }
         else if(elbow==ON)
         {
                    if (sign==NEG)
                                                       el_pos+=increment;
                    else el_pos-=increment;

                el_p[j]=el_pos;
                j++;
            }
         else if(wrist==ON)
            {
                    if (sign==NEG)
                                                       wr_pos+=increment;
                    else wr_pos-=increment;

                wr_p[k]=wr_pos;
                k++;
                }
         data_pre=data_now;
         }
/*************************************************/
            err_sh()        /* error calculations. */
            {
                    return(des_sh_pos-sh_pos);

            }
            err_el()
            {
                    return(des_el_pos-el_pos);

            }
            err_wr()
            {
                    return( des_wr_pos-wr_pos);

            }
/***************************************************************/
            output(comm)  /* send the command to the p.s.*/
            int comm;
            {
                int out;
                if (shoulder==ON)
                    comm= - comm;
                if( comm < 0)
                    {
                    s_neg_sgn();
```

```
                    out=(63-comm ) | 192;
    /* check the transmitter buffer before sending more data. */
    tx_buf_free();
                outp(srl_data,out);
                tx_buf_free();
                out-=128;
                outp(srl_data,out);

                }
                else  {
                s_pos_sgn();
                        out= (63-comm) | 192;
                tx_buf_free();
                outp(srl_data,out);
                tx_buf_free();
                out-=128;
                outp(srl_data,out);
                }
        }
```

/**/

```
        s_relay1()    /* set the shoulder relay on. */
        {
            tx_buf_free();
          outp(srl_data, 160);
          tx_buf_free();
          outp(srl_data, NOP);
          }
        s_relay2()    /* set the elbow relay on. */
        {
          tx_buf_free();
          outp(srl_data, 176);
          tx_buf_free();
          outp(srl_data, NOP);
          }
        s_relay3()    /* set the wrist relay on. */
        {
          tx_buf_free();
          outp(srl_data, 168);
          tx_buf_free();
          outp(srl_data, NOP);
          }
        s_pos_sgn()    /* set the p.s. supply positive. */
         {
          tx_buf_free();
          outp(srl_data, 152);
          tx_buf_free();
```

Chap. 14   Fuzzy Control of Robotic Manipulator   325

```
            outp(srl_data, NOP);
            }
            s_neg_sgn()    /* set the p.s. negative. */
            {
            tx_buf_free();
            outp(srl_data, 144);
            tx_buf_free();
            outp(srl_data, NOP);
              }

               s_0volts()    /* turn off the p.s. */
               {

               tx_buf_free();
               outp(srl_data,0xff);
               tx_buf_free();
               outp(srl_data,127);
               }
```

/***/

```
        ini_zero()   /* iniatialize the position to zero. */
        {
              sh_pos=0;
              el_pos=0;
              wr_pos=0;
              }
        upd_last()  /* update the positions from the last run. */
          {
             FILE *in;
             int ch;

             if ((in=fopen("sto_rhn.dat", "r")) != NULL)
                 {
                    sh_pos=getc(in);
                    el_pos=getc(in);
                    wr_pos=getc(in);
                 }
                 {
                 printf(" The file does not exist.\n");
                 printf(" I think you should initialize the Rhino.\n");
                 }
                 fclose(in);
          }
```

/***/

326   Chap. 14   Fuzzy Control of Robotic Manipulator

```
exit1() /* exit after storing the present position in a file. */
{
    int l;
    FILE *out,*out2;

    out=fopen("rhn_last","w");
    out2=fopen("traj.m","w");
    fprintf(out,"%d %d %d\n",sh_pos, el_pos, wr_pos );
    fprintf(out2,"sh=[\n");
    for(l=0;l<i;l++)
    fprintf(out2,"%d \n",sh_p[l]);
    fprintf(out2,"];\n\n");

    fprintf(out2,"el=[\n";
    for(l=0;l<j;l++)
    fprintf(out2,"%d \n",el_p[l]);
    fprintf(out2,"];\n\n");

    fprintf(out2,"wr=[\n");
    for(l=0;l<k;l++)
    fprintf(out2,"%d \n",wr_p[l]);
    fprintf(out2,"];\n\n");

    fclose(out);
    fclose(out2);
    exit(0);
}

exit2()
{
    FILE *out;
    s_0volts();
    out=fopen("store","w");
    fprintf(out,"%d %d %d",sh_pos, el_pos, wr_pos );
    fclose(out);
    printf("ABNORMAL TERMINATION OF THE PROGRAM.\n");
    exit(1);
}
```

/***/

/* Routine to set up the baud rate and data format for interfacing
with the Rhino robot through the RS232c communication port.   */

/* set the data having 8 bits, 2 stop bits, no parity.   */ set_data()

```
    {
       outp(0x3fb,0x07);
    }
/*              set the baud rate to 9600           */
         set_baud()
  {
    int tmp;
    tmp = get_data() | 0x80 ;
    outp(0x3fb, tmp);
           outp(0x3f8, 0x0c);
    outp(0x3f9, 0x00);
    tmp = tmp & 0x7f ;
    outp(0x3fb, tmp);
         }


int get_data()
   {
         return ( inp(0x3fb));
         }



  int get_stat()
  {
         return(inp(0x3fd));
         }



/* s_xmitstatus function returns the status of the transmitter buffer. */
int       s_xmitstat()
{
   return ( inp(0x3fd) & 0x20);
   }

  /* s_rcvstat function returns the status of the  data ready. */
  int s_rcvstat()
{
   return ( inp(0x3fd) & 1);
        }




         tx_buf_free()

{
         long int tim = 29999;
     int status;
     do
```

```
          {
        tim--;
        status = s_xmitstat();
          }while (status !=32 && tim !=0);
              if (tim == 0)
        contErr();
        /* communication to the controller lost. */
          }

              contErr()
              {
  printf("\nCOMMUNICATION TO THE CONTROLLER LOST... ABORT.\n"); }
        /* A small delay function. */
              Delay()
      {
              long int n= 59999;
        long int p=0;
        while(p++ < n);
        }
        /* Sampling delay. */
              sample_delay()
      {
              long int n= 6999;
        long int p=0;
        while(p++ < n);
  }

/*************************************************************/
```

15

USE OF FUZZY LOGIC CONTROL IN ELECTRICAL POWER GENERATION

Erlendur Kristjánsson
Denis Barak
Kenneth Plummer
University of New Mexico

In this chapter, the application of a real-time fuzzy controller is presented in order to improve the power system stability. This is done with the means of the fuzzy membership functions whose values are computed depending on the measured generator frequency and voltage. The controller was tested in real time for different loads on the generator.

15.1   INTRODUCTION

Currently, improvements in the dynamics of power systems are made with a help of lead-lag, robust, or self-tuning power system stabilizers (PSS). The lead-lag PSS is the most widely employed in power systems because of its simplicity. In this case, a fixed gain controller is designed and adjusted to certain operating conditions. Since the operating point of a power system changes as a result of continuous load changes or parameter drift, robust and self-tuning PSS can be implemented. Performance of the adaptive PSS is much better than fixed parameter PSS, but the cost is significantly increased because of employment of powerful computers, which are necessary to achieve real-time model identification.

In the last few years, several papers, [1-5], proposed the idea for the use of fuzzy logic in power systems. From those, [1] and [2] can be closely related toward the simplification of design and application of robust and self-tuning PSS. In this case, fuzzy logic gives the ability to design an "adaptive" PSS without the knowledge of the system model. "Adaptive," in this case, means that the stabilizer is immune to the changes in the operating point but, at the same time, expensive equipment is not necessary because real-time model identification is not performed.

In this chapter, unlike [1] and [2], a fuzzy logic PSS is proposed for real-time applications.

15.2   SYSTEM SET-UP

The system consists of the following units:

1. DC-motor, (prime mover)
2. AC-generator, (synchronous machine)
3. Fuzzy logic controller
4. Control circuits and measurement circuits

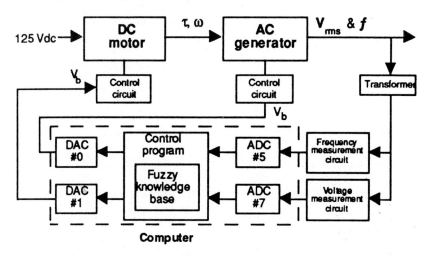

Figure 15-1. Block diagram of system set-up.

To simulate the prime mover/generator concept, two electro-mechanical machines were chosen: one to act as an externally driven motor, whose job is to mechanically rotate a shaft, and another to serve as an AC generator (also called an

Chap. 15 Fuzzy Logic Power Generation Control

alternator), providing three-phase voltage in the output. The goal is to approximate a power generation station as closely as possible. Figure 15-1 shows a block diagram of the system set-up.

The Prime Mover

In power generation, the machine designed to perform the shaft rotation is actually a turbine, typically driven by rising steam. This turbine system is referred to as the prime mover. The vaporized water is produced by heat from the combustion of natural gas, fossil fuels such as oil or coal, or by nuclear power. The type of fuel used to ultimately drive the shaft greatly depends upon the natural resources found in the area of the load requirement. For this reason, high voltage power lines connect remote generators to electrical load centers.

However, for the purpose of this experiment and with access to high voltage supplies, a Hampden DM-100 DC machine was chosen as the prime mover. As shown in Figure 15-2, the machine is wound with both a shunt field winding (located on the machine's stator component) and a series field winding (located on the machine's rotor component). Thus, the machine is a compound DC machine capable of operating as a motor or as a generator.

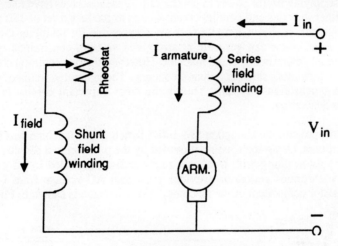

Figure 15-2. DC-motor with compound winding.

The operation of the machine is fairly simple. With a single applied input voltage of 125 V_{DC}, an armature current and a field current are created. This set-up with the non-separate excitation of the field and armature windings was a major concern in the choice of the compound machine, along with its flexibility and common configuration. At speeds near zero and for starting purposes, the shunt field current should be held high to obtain maximum torque, following from equation (1):

$$T = k_a \Phi_d i_a \qquad (1)$$

Here, Φ_d is the flux of the "direct" axis due to the magnetic field generated by the stator and is directly proportional to the shunt field current.

As the motor increases speed, the resistance of the rheostat is increased, forcing more current through the armature winding and less current through the field winding. From equation (2),

$$\omega_m = \left[\frac{E_a}{k_a}\right] \Phi_d^{-1} \qquad (2)$$

where ω_m is the mechanical speed (speed of rotor, therefore speed of shaft), the rotor speed responds as inversely proportional to field current, for Φ_d defined as before. Therefore, for a decrease in field current (or an increase in armature current), the rotor speed will increase proportionally.

The Synchronous Generator

At power generation stations, large three-phase synchronous generators (typically many millions of Watts) are locked in parallel with a network of busses, lines, and transformers, supplying useful power to purchasers. These machines have many poles and are wired together in series and parallel combinations to produce a set of 120 degree out-of-phase voltage curves. Since transformers are responsible for sculpting the necessary peak voltage, a generator can generate at any given voltage, constrained only by the current and overall power rating of the machine. However, the one common characteristic that all parallel generators share is output frequency. The generated frequency is vital to maintaining a synchronous network. This is the most important control procedure in power systems generation.

For this project, the Hampden SM-100-3 Synchronous machine was selected for generation purposes. Using the rotation provided by the prime mover directly coupled to the alternator's rotor, along with a magnetization current provided by an external DC source, the synchronous generator provides an output AC voltage from its armature connections (stator component of the machine). The schematic is shown in Figure 15-3.

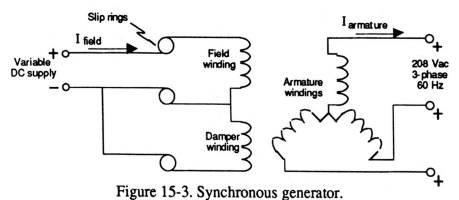

Figure 15-3. Synchronous generator.

Similar to DC machine operation, changes in the field current effect the synchronous machine's output characteristics. With the prime mover being the dominant force behind the rotor speed, the frequency mostly can be contributed to the rotational speed of the coupled shaft. Therefore, the DC field current of the alternator mainly contributes to the voltage level generated in the armature (which may differ slightly from actual output voltage, due to winding losses), as shown by equation (3):

$$E_{af} \text{ (rms)} = \left[\frac{\omega L_{af}}{\sqrt{2}}\right] I_f \qquad (3)$$

where E_{af} is the armature voltage, L_{af} is the mutual inductance between the field winding and winding "a" of the armature, and ω is the synchronous speed.

Fuzzy Logic Controller

The fuzzy logic controller (FLC) is the heart of the system; it makes logic decisions depending on the input/output behavior of the generating unit. It can be seen from Figure 15-1 that the output of the generator is fed via the measurement circuits and an analog-to-digital converter (ADC) into the computer, and that the generator is controlled via a digital-to-analog converter (DAC) and control circuits. Therefore, the FLC is a software program (Appendix B) that interacts with the generator such that the output of the generator is maintained at 60 Hz and 120 V_{rms}.

Note that the system is stand-alone and its output is not effected by a stiff network as it would if it were a part of a power system. For this reason, any load change will have a big effect on both the output frequency and the output voltage. The FLC will, therefore, move to different operating points, stabilizing (as fast as possible) the output of the generator to the desired values.

Control Circuits

The control of the motor-generator pair can be accomplished only through the use of circuits external to the fuzzy logic controller (FLC), i.e., circuits that interface the FLC with the machines, and circuits that take information from the machines to FLC. Since the field circuits of both machines are to be used for controlling frequency and output voltage level, the external circuits tied to the machines must be capable of handling relatively high power and current. They must also exhibit good heat dissipation characteristics to avoid nonlinear behavior over extended operation periods. Also, since the input voltage of the ADC is limited to 0-5 V_{DC}, the output generator voltage must be altered to be used in the computer.

Replacing the bulky, inaccurate field rheostat in Figure 15-2 is a simple NPN transistor circuit shown in Figure 15-4. Here, collector current can be controlled fairly accurately with the base current [equation (4)], which linearly changes with the base voltage change. Though the transistor gain (β) can vary in behavior as a function of temperature, a proper heat sink can help protect the circuit for reasonable periods of time.

Of course, other means of current control can be more effective and completely independent of intrinsic transistor gain.

$$I_{collector} = \beta \cdot I_{base} \qquad (4)$$

The base voltage is changed through DAC#1 and the op-amp circuit in Figure B of Appendix A. Therefore, by increasing voltage from DAC#1, the transistor current is increased (I_{field} in Figure 15-2), decreasing $I_{armature}$ and, consequently, decreasing the speed of the prime mover. Decreasing the output voltage of DAC#1 has an opposite effect [equations (1) and (2)]. Thus, for starting, the base current should be maximum, decrease to steady state values (therefore speeding up the rotor), and vary positively or negatively to control the rotor speed.

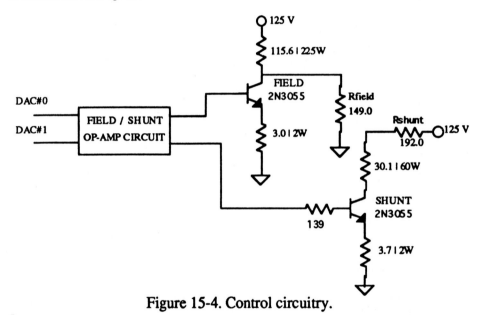

Figure 15-4. Control circuitry.

For the magnetization field of the AC generator, a similar NPN transistor circuit is used as the one mentioned above. The only difference is that DAC#0 is used to control the collector voltage level between 50 and 60 V_{DC} (see Figure 15-4). This circuit replaces the variable DC supply in Figure 15-3; by increasing the output of DAC#0, transistor current is increased so that collector voltage is decreased, consequently decreasing the output voltage of the generator, and vice versa [equation (3)].

Measurement Circuits

To achieve desired input levels to the ADC (0-5 V_{DC}), the alternator's output must be converted from 120 V_{rms} to 5.58 V_{rms} by means of a transformer. For a voltage measurement circuit, the transformer output voltage is passed through an RMS to DC converter and then scaled through op-amp circuit shown in Figure A of Appendix A.

Chap. 15 Fuzzy Logic Power Generation Control 335

Therefore, the alternator voltage of $120\pm20/\sqrt{3}$ V_{rms} is converted and to 0.172-4.816 V_{DC} in the input to the computer. A diagram of the measurement circuitry is shown in Figure 15-5.

For the frequency measurement circuit, the transformer voltage is rectified and filtered, and a square pulse is created by means of a comparator with the same frequency as the output of the generator. This square pulse is passed through a frequency to voltage converter, and the output DC voltage of the converter is scaled with the op-amp circuit from Figure A of Appendix A. Therefore, the alternator frequency between 55-65 Hz is converted and scaled to 0.172-4.816 V_{DC} in the input to the computer. The diagram of this circuit is also shown in Figure 15-5.

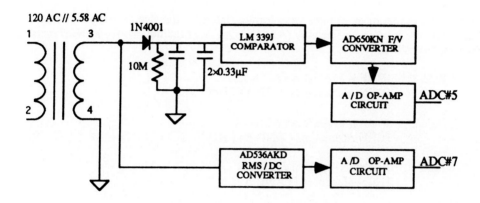

Figure 15-5. Measurement circuitry.

15.3 FUZZY KNOWLEDGE BASE

To be able to form a knowledge base the input and output signals had to be given a working range. It was decided that the voltage could fluctuate in a range between $110V_{rms}$ and $130V_{rms}$, and the frequency in a range between 55 Hz and 65 Hz. These ranges were then divided into sets, and each set was given membership functions.

The next step was to form a knowledge base. But before that was possible, measurement of the input/output behavior of the system was needed. From these measurements the fuzzy rules were created. Here is an example of making of a fuzzy rule:

– Changing load from infinity to 1000Ω:

Frequency: 60Hz → 57Hz ⇒ ΔF = 3Hz
Voltage: $120V_{AC}$ → $113V_{AC}$ ⇒ $\Delta V_{AC} = 7V_{AC}$

– Control actions to get frequency back to 60Hz and voltage to 120V$_{AC}$:

I_{shunt}: 0.310V$_{DC}$ → 0.261V$_{DC}$ ⇒ ΔI_{shunt} = –0.049A$_{DC}$
V_f : 54.35V$_{DC}$ → 56.8V$_{DC}$ ⇒ ΔV_f = 2.45V$_{DC}$

– Then the following rule is generated from the fuzzy sets:

IF ΔF is L AND ΔVAC is M
  THEN $\Delta Ishunt$ is N AND ΔVf is P

where L stands for the membership function Low, M for Medium, N for Negative, and P for Positive.

In this way, the entire set of rules were formed, based on the change in input and output. The final rule base is as follows. (Ab. L=Low, M=Medium, H=High, N=Negative, Z=Zero, P=Positive)

Rule 1:   IF ΔF is L AND ΔVAC is L
          THEN $\Delta Ishunt$ is N AND ΔVf is N

Rule2:   IF ΔF is L AND ΔVAC is M
          THEN $\Delta Ishunt$ is N AND ΔVf is Z

Rule3:   IF ΔF is L AND ΔVAC is H
          THEN $\Delta Ishunt$ is N AND ΔVf is P

Rule4:   IF ΔF is M AND ΔVAC is L
          THEN $\Delta Ishunt$ is Z AND ΔVf is N

Rule5:   IF ΔF is M AND ΔVAC is M
          THEN $\Delta Ishunt$ is Z AND ΔVf is Z

Rule6:   IF ΔF is M AND ΔVAC is H
          THEN $\Delta Ishunt$ is Z AND ΔVf is P

Rule7:   IF ΔF is H AND ΔVAC is L
          THEN $\Delta Ishunt$ is P AND ΔVf is N

Rule8:   IF ΔF is H AND ΔVAC is M
          THEN $\Delta Ishunt$ is P AND ΔVf is Z

Rule9:   IF ΔF is H AND ΔVAC is H
          THEN $\Delta Ishunt$ is P AND ΔVf is P

and the membership functions are shown in Figures 15-6, 15-7 and 15-8.

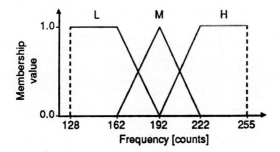

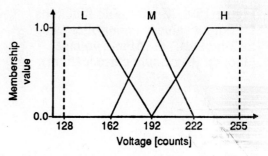

Figure 15-6. Membership functions for input values of controller. (a) Frequency (counts), above; (b) Voltage (counts), right.

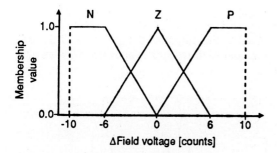

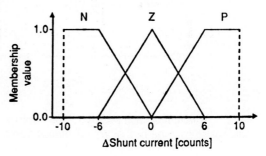

Figure 15-7. Membership functions for output values of controller using an 8-bit DAC. (a) Field voltage (above); (b) ΔShunt current (right).

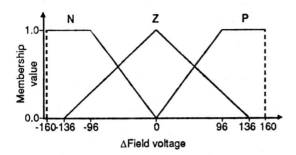
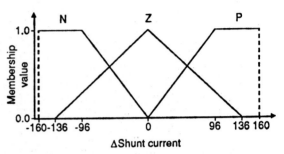

Figure 15-8. Membership functions for output values of controller using a 12-bit DAC. (a) ΔField voltage (above); (b) ΔShunt current (right).

The fuzzy sets, their membership functions, and the fuzzy rules were then assembled into a fuzzy knowledge base by using the Togai InfraLogic (TIL) Shell in Microsoft Windows. The output file of the TIL Shell (Appendix C) was then compiled to generate a C-subroutine, which was then linked to the main control program.

15.4 EXPERIMENT

To test the system, a load unit was connected to the output of the generator. This load unit is a box of parallel connected resistors, and each resistor is activated by a switch. Figure 15-9 shows the load unit diagram.

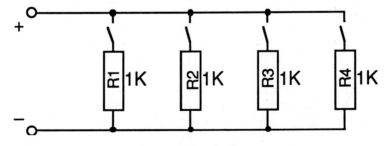

Figure 15-9. Load unit.

Note that each additional load does not add resistance on the output but decreases the resistance and in that way demands power. Therefore the generator must increase its

Chap. 15 Fuzzy Logic Power Generation Control

power output, which requires more power to the DC-motor. Results in the following section were obtained by switching two load resistors at the same time in the following sequence:

1) NO LOAD
2) R1-R2 ON
3) R3-R4 ON
4) R3-R4 OFF
5) R1-R2 OFF

The loads were turned on/off at the same time points.

15.5 RESULTS

The frequency and voltage responses are shown in Figures 10 and 11, respectively. In Figure 15-10 (8-bit case), it can be seen that signals settle around their optimal values but their fluctuation is too big to be satisfactory. The problem is in the sensitivity of the control signals, shown in Figures 12 and 13 (8-bit case); i.e., even though the system outputs are still not completely settled, the control signals have already settled because change in the input is too small to achieve any change in the output. Therefore, the 8-bit DAC has insufficient precision for this particular circuit.

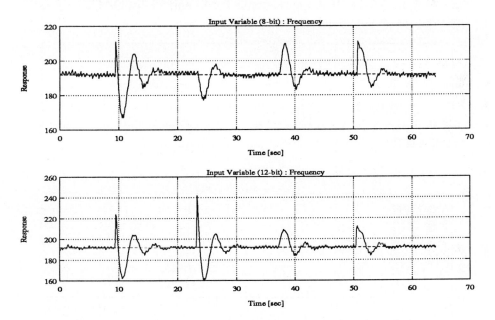

Figure 15-10. Frequency response for 8- and 12-bit DAC.

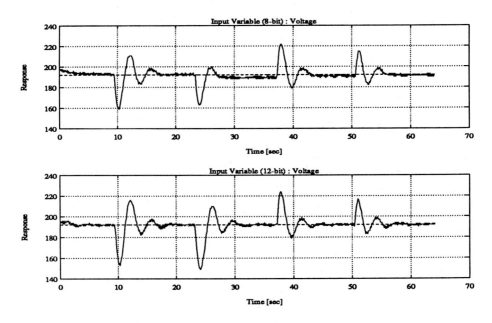

Figure 15-11. Voltage response for 8- and 12-bit DAC.

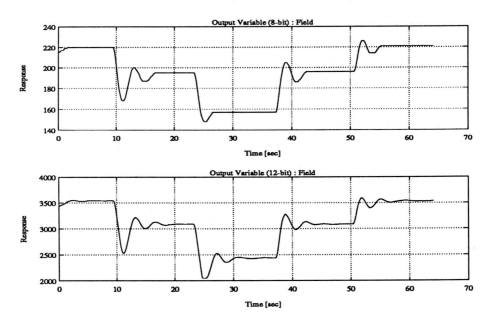

Figure 15-12. Control signal from 8- and 12-bit DAC for field voltage control.

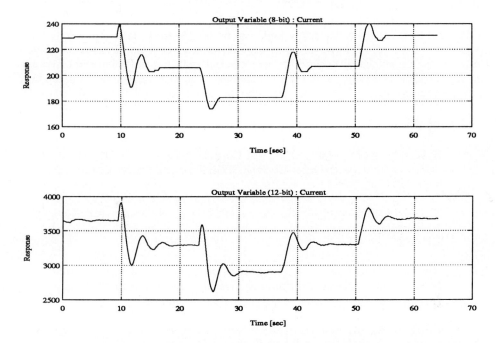

Figure 15-13. Control signal from 8- and 12-bit DAC for shunt current control.

To solve this problem, the DAC bits are increased to 12, giving finer control resolution. In Figures 10 and 11 (12-bit case), the results of the 12-bit controller are shown. As seen in Figure 15-10, the 12-bit case, the fluctuations around optimal values were significantly reduced, which supports the use of a more expensive DAC. Furthermore, the 12-bit DAC achieved settling of frequency and voltage with zero offset, as opposed to its 8-bit counterpart (see Figures 8 and 9).

However, in Figure 15-10, there is an apparent inconsistency in the second load disturbance. The 8-bit controller seems to suppress the transient response better than the 12-bit controller. This can be a misleading interpretation because the controller, at the moment of disturbance, was already executing rules to control steady state oscillations. Referring to Figure 15-11, it can be seen that the control value at that time is changing in the opposite direction from that which is needed to suppress the new disturbance. Another reason for this inconsistency is the inherently large time constants in the system. These characteristics achieve relatively slow response times given sudden changes in the control signals. Therefore, in this case, these two system characteristics caused this abnormally large overshot to occur.

15.5 CONCLUSION

In this chapter it is shown that fuzzy logic can be used with success in power systems control. The main result, however, is that this method gives a simple way to design a controller for a nonlinear system without complicated real-time modeling and expensive hardware. Future studies of this control scheme would be to develop a model of a power system, and compare this control method with other non-fuzzy control methods, as in [6].

REFERENCES

[1] Hassan, M. A. M., Malik, O. P., and Hope, G. S., *A Fuzzy Logic Based Stabilizer for a Synchronous Machine*, IEEE Trans. on Energy Conversion, Sept. 1991, pp. 407-413.

[2] Hsu, Y.-Y., and Cheng, C.-H., *Design of Fuzzy Power System Stabilisers for Mulitmachine Power Systems*, IEE Proceedings-C, May 1990, Vol. 137, pp. 233-238.

[3] Hiyama, T., and Lim, C. M., *Application of Fuzzy Logic Control Scheme for Stability Enhancement of a Power System*, IFAC Symposium on Power Systems and Power Plant Control, Aug. 1989, Singapore.

[4] David, A. K., and Rongda, Z., *An Expert System with Fuzzy Sets for Optimal Planning*, IEEE Trans. on Power Systems, Feb. 1991, pp. 59-65.

[5] Tomsovic, K, *A Fuzzy Linear Programming Approach to the Reactive Power/Voltage Control Problem*, IEEE Trans. on Power Systems, Feb. 1992, pp. 287-293.

[6] Yousef, H., and Simaan, M. A., *Model Reference Adaptive Control of Large Scale Systems with Application to Power Systems*, IEE Proceedings-D, July 1991, Vol. 138, pp. 321-326.

Chap. 15 Fuzzy Logic Power Generation Control 343

APPENDIX A

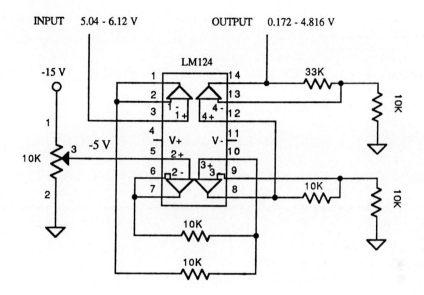

Figure A. A/D op-amp circuit.

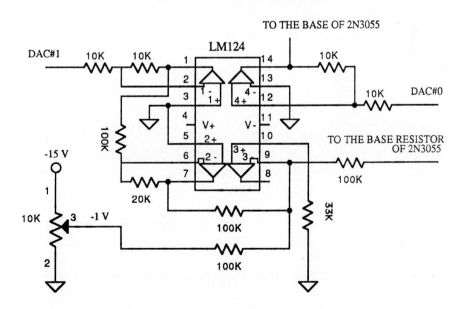

Figure B. Shunt/Field op-amp circuit.

APPENDIX B

```c
/*******************************************************
 *
 *   SUBJECT  : FUZZY CONTROLLER FOR POWER GENERATOR
 *
 *   PROGRAM  : POWGEN.C
 *
 *   AUTHORS  : DENIS BARAK &  ERLENDUR KRISTJÁNSSON
 *
 *   HARDWARE : KEITHLEY METRABYTE DAS-4 & DAC-02
 *
 *******************************************************/

#include "constant.h"
#include "routines.h"
#include "pow_ctrl.h"
#include <sys/types.h>
#include <sys/timeb.h>
#include <stdio.h>
#include <time.h>

main ()
{

   /* Initialize variables ..... */
   signed short int
      delta_field, delta_current, field,
      current, frequency, voltage;
   int          rules[22];

   /* Crule sampling frequency, but was use because
of inoperative interrupt circuit on ADC-board.
   */
   int          sampling_count=30000;

   long    i=0, j=0, k=0;
   FILE    *fp1, *fp2;
```

Chap. 15 Fuzzy Logic Power Generation Control

```c
  /* Clear rule vector */
  for(i=0; i<9; i++)
     rules[i] = 0;

  /* Open the files for data output ... */
  fp1 = fopen ("inout12.dat", "w");
  fp2 = fopen ("fuzzyout.dat", "w");

  /* Clear the screen and output initial data
...*/
  clear_screen();
  dtoa_12 (start_current_12, dtoa1_loc);

  /* Waiting for user to start the controller ...
*/
  printf("Hit any key to start the control
process!!! \n\n");
  do
  {
  }while (!kbhit());
  getch();

  /* Output initial data ... */
  dtoa_12 (initial_field_12, dtoa0_loc);
  field = initial_field_12;
  dtoa_12 (initial_current_12, dtoa1_loc);
  current = initial_current_12;

  printf("Hit any key again!!! \n\n");
  do
  {
  }while (!kbhit());
  getch();

  printf("Control process is in the progress!!!
\n\n");

  /* Implement the controller loop ... */
  do
  {
```

```c
    /* Set up A/D in analog mode, and read frequency
from CH#5 */
    frequency = atod_analog
        (enable_5_ints, disable_5_ints);
    if (frequency < 128)     frequency = 128;

    /* Set up A/D in analog mode, and read line
voltage from CH#7 ..... */
    voltage = atod_analog
        (enable_7_ints, disable_7_ints);
    if (voltage < 128)    voltage = 128;

    /* Get I/O from the fuzzy controller ..... */
    power_controller (frequency, voltage,
        &delta_current, &delta_field);

    /* Output field voltage to the generator through
D/A#0   */
    field += delta_field;
    if (field < 2048)    field = 2048;
    if (field > 4095)    field = 4095;
    dtoa_12 (field, dtoa0_loc);

    /* Output shunt current to the DC motor through
D/A#1    */
    current += delta_current;
    if (current < 2048)   current = 2048;
    if (current > 4095)   current = 4095;
    dtoa_12 (current, dtoa1_loc);

    /* Print information to the file each sampling
time ... */
    if(j++ == 0)
    {
        fprintf(fp1,"%d %d %d %d \n",
            frequency, voltage, field, current);
        fprintf(fp2,"%d %d %d %d \n", frequency,
            voltage,delta_field, delta_current);
    }
    if(j==1)  j=0;
```

Chap. 15 Fuzzy Logic Power Generation Control

```
    /* Counter loop representing real-time interrupt
... */
    i=0;
    do
    {
        i++;
    }while(i<sampling_count);

  }while (!kbhit());

  /* Close files for data output ..... */
  fclose (fp1);
  fclose (fp2);

}

/******************************************************
*   End of the file.
******************************************************/


/***************** routines.c ****************/

#include "routines.h"
#include "constant.h"

/*************** clear_screen ***************/

void clear_screen (void)
{

    int i;

    for (i=1; i<=24; i++)
        printf("\n");

}

/************** end clear_screen ***************/
```

```c
/*************** dtoa - 12 bits ****************/

void dtoa_12 (int vout, int dtoa_12_loc)
{
      /*  Initialize variables ..... */
      unsigned int value;

      /* Output voltage through D/A ..... */
      value = (vout << shift_dtoa_output);
      outpw(dtoa_12_loc, value);
}

/************* end dtoa - 12 bits **************/


/**************** dtoa - 8 bits ****************/

void dtoa_8 (int vout, int dtoa_8_loc)
{
      /*  Initialize variables ..... */
      unsigned int value;
      value = vout;

      /* Output voltage through D/A ..... */
      outp(dtoa_8_loc, 0);
      outp(dtoa_8_loc+1, value);
}

/************** end dtoa - 8 bits **************/
/***************** atod_analog *****************/

int atod_analog (int enable_ints, int disable_ints)
{
      /*  Initialize variables ..... */
      char status;
      int data;

      /* Set up to read a/d channel with interrupt
            enabled ..... */
      outp(atod_ctrl, enable_ints);
```

Chap. 15 Fuzzy Logic Power Generation Control

```c
        /* Wait for the interrupt signal (interrupt
            when D3=1) ... */
/*      while( (status=inp(atod_stat) &
masking_interrupt) == 0)   */

        /* Clear the interrupt ..... */
        outp(atod_ctrl, disable_ints);

        /* Start a/d conversion ..... */
        outp(atod_strt,1);

        /* Wait until conversion complete ..... */
        while((status=inp(atod_stat)&masking_
            end_of_conversion)!=0);

        /* Read a/d data and return it to the
            main program ..... */
        data = inp(atod_data);

        return (data);
}

/************** end atod_analog ***************/

/*************** end routines.c ***************/


/****************** routines.h ****************/


void    dtoa_12 (int, int);
        /* 12-bit D/A output operation.  The
            first number represents the desired
            output voltage, while the second number
            represents D/A's channel address        */

void    dtoa_8 (int, int);
        /* 8-bit D/A output operation.  The
            first number represents the desired
            output voltage, while the second number
            represents D/A's channel address        */
```

```c
int   atod_analog (int, int);
      /* These numbers represent enable and
         disable combinations for certain
         channel on A/D respectively;the return
         value from this subroutine is "data"
         and represents the input value from
         certain channel        */

void clear_screen (void);


/*************** end routines.h ****************/


/***** constant.h -- constants and addresses *****/


#include <conio.h>
#include <stdlib.h>
#include <stdio.h>
#include "tilcomp.h"


#define atod_strt              0x0200
#define atod_data              0x0201
#define atod_ctrl              0x0202
#define atod_stat              0x0203

#define dtoa0_loc              0x0208
#define dtoa1_loc              0x020a


#define zero_atod              127
#define zero_dtoa_8            127
#define zero_dtoa_12           2068
#define shift_dtoa_output      4


#define enable_5_ints          0x0d
#define disable_5_ints         0x05
```

```c
#define enable_7_ints                   0x0f
#define disable_7_ints                  0x07
#define masking_interrupt               0x08
#define masking_end_of_conversion       0x80
#define masking_digital_input           0x30

#define start_current_8                 255
#define initial_field_8                 214
#define initial_current_8               229
#define start_current_12                4095
#define initial_field_12                3424
#define initial_current_12              3664

#define hex_to_decimal_coeff            16
#define percent                         100.0
#define t_sample                        0.05
#define voltage_conversion              240.4
#define atod_dtoa_conversion            12.67

/************** end of constant.h ***************/
```

APPENDIX C

```
PROJECT power_controller

    VAR Frequency

        TYPE signed word
        MIN 128
        MAX 255

        MEMBER LOW

            POINTS 128,1 162,1 192,0
        END

        MEMBER MED

            POINTS 162,0 192,1 222,0
        END

        MEMBER HIGH

            POINTS 192,0 222,1 255,1
        END
    END

    VAR Voltage

        TYPE signed word
        MIN 128
        MAX 255

        MEMBER LOW

            POINTS 128,1 152,1 192,0
        END

        MEMBER MED

            POINTS 162,0 191.5,1 222,0
        END
```

Chap. 15 Fuzzy Logic Power Generation Control

```
        MEMBER HIGH

            POINTS 192,0 233,1 255,1
        END
    END


VAR Field

    TYPE signed word
    MIN -160
    MAX 160

    MEMBER NEG

        POINTS -160,1 -94,1 0,0
    END

    MEMBER ZO

        POINTS -136,0 0,1 136,0
    END

    MEMBER POS

        POINTS 0,0 96,1 160,1
    END
END

VAR Current

    TYPE signed word
    MIN -160
    MAX 160

    MEMBER NEG

        POINTS -160,1 -96,1 0,0
    END
```

```
    MEMBER ZO

        POINTS -136,0 0,1 136,0
    END

    MEMBER POS

        POINTS 0,0 96,1 160,1
    END
END


FUZZY Power_Rules

    RULE Rule0000

        IF  (Frequency IS LOW)
    AND (Voltage IS LOW) THEN
            Current=NEG
            Field=NEG
    END

    RULE Rule0001

        IF  (Frequency IS LOW)
    AND (Voltage IS MED) THEN
            Current=NEG
            Field=ZO
    END

    RULE Rule0002

        IF  (Frequency IS LOW)
    AND (Voltage IS HIGH) THEN
            Current=NEG
            Field=POS
    END

    RULE Rule0003
```

```
        IF (Frequency IS MED)
AND (Voltage IS LOW) THEN
        Current=ZO
        Field=NEG
END

RULE Rule0004

        IF (Frequency IS MED)
AND (Voltage IS MED) THEN
        Current=ZO
        Field=ZO
END

RULE Rule0005

        IF (Frequency IS MED)
AND (Voltage IS HIGH) THEN
        Current=ZO
        Field=POS
END



RULE Rule0006

        IF (Frequency IS HIGH)
AND (Voltage IS LOW) THEN
        Current=POS
        Field=NEG
END

RULE Rule0007

        IF (Frequency IS HIGH)
AND (Voltage IS MED) THEN
        Current=POS
        Field=ZO
END
```

```
        RULE Rule0008

            IF  (Frequency IS HIGH)
        AND (Voltage IS HIGH) THEN
                Current=POS
                Field=POS
        END
    END

    CONNECT
        FROM Frequency
        TO Power_Rules
    END

    CONNECT
        FROM Voltage
        TO Power_Rules
    END

    CONNECT
        FROM Power_Rules
        TO Field
    END

    CONNECT
        FROM Power_Rules
        TO Current
    END
END
```

16

FUZZY LOGIC CONTROL OF RESIN CURING

Mark K. McCullough
University of New Mexico

Plastics of various types are found in almost all major products of the 20th Century. From ball point pens to jet aircraft engines, plastic is replacing metal parts for cost reductions and weight reductions. However, complex parts such as those found in the aircraft industry do not necessarily have a manufacturing cost reduction with the implementation of some plastic resin systems. On the contrary, a need for an exotic resin system's material characteristics can require a parts designer to pay a premium for a plastic part. This paper addresses the problem of cost effective manufacturing of plastic resin parts by utilizing a fuzzy logic controller. Experimentation is done with off-line control simulations.

16.1   INTRODUCTION

Processing

For general plastic manufacturing, process variables such as temperature and pressure are correlated to a time scale that makes up what manufacturing engineers call the process window. This process window is linked to the resins' viscosity during a process cycle. Normally, pressure is applied when the resin is just starting to harden (See Figure 16-1).

To increase product manufacturing rates and yields, new sensors are being experimented with to monitor the plastics' chemical reactions. By monitoring chemical reactions to process variables, the process can be driven to an optimization point within the process window. In certain complex plastic resin systems, achieving the optimization point is not only very difficult due to different raw material characteristics, but also critical for achieving the materials' design criteria.

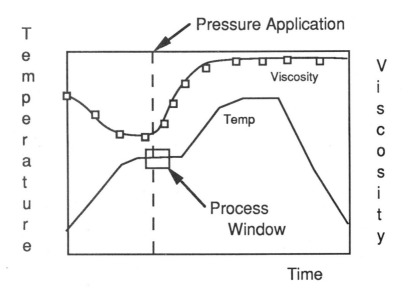

Figure 16-1. Resin is forced through a temperature profile. Pressure has to be applied during a process window to obtain certain material properties.

For example, pressure that is applied early in a cure cycle can force the viscous resin to exit the system resulting in a high fiber to resin ratio in a composite part utilizing a fiber matrix. Pressure applied late in the cycle cannot compact the material due to the resin's stiffened state. Both early and late pressure application results in a mechanically weakened composite structure. Given a standard temperature profile, the time at which pressure should be applied is somewhat elusive due to;

- Manufacturers' supplied resin has batch to batch variations due to minor modifications in processing.
- Different manufacturers' supplying slightly different resin systems.
- Time material has been outside the storage state of zero degrees Fahrenheit.
- Number of times the material has been thawed can effect processing.
- Process being driven faster to increase through-put will shrink the size of the process window.

Chap. 16 Fuzzy Logic Control of Resin Curing

Fuzzy Control

Varying methods of control can be formulated to solve the problem of when to apply pressure to a resin system given the input data. A math model could provide a basis for control or data gathering experiments could provide a data base for approximating output control. However, the cost of modeling complex resin systems and data base experiments can be high. Even if a process model can be developed for control, there is a high degree of probability that the model will have to change as the resin is modified for improvements. Approximating a math model with experimental data points may not be as costly as modeling but, practical implementation limits the number of data sets. This limit can lead to interpolation errors and/or calculation speed problems. Implementing Fuzzy Logic Control over a resin process provides the ability to associate the normal process variables of time and temperature for a pressure signal, and also implement data relevant to the material's chemical reaction as the resin changes without a real world model. With a Fuzzy Logic Controller, the measured process parameters can be associated with appropriate membership ranges that can produce a meaningful output. Instead of the controller looking for a specific value it would give confidence to all values in a specified range until the membership output is high enough to trigger a reaction.

16.2 TECHNICAL APPROACH

Employ a dielectric sensor and impedance analyzer to sense a resin system's degree of chemical reaction. Most resin systems are fairly high in free ions that can be sensed by the dielectric sensor. As heat is applied and the resin starts to cross-link, the free ions are used up in chemical reactions or immobilized. Therefore, by monitoring ionic resonance, a reproducible point of pressure can be found that is dependent on the chemical stage of the resin instead of time and temperature profile alone. A fuzzy logic algorithm will be used to monitor temperature, time and the capacitive loss factor in order to provide a crisp output point for pressure control.

16.3 EXPERIMENTAL DATA

Figure 16-2 is a sketch representing the equipment used for the experiments. An MS DOS computer is used to initialize the data acquisition system for frequency of the excitation signal and storage of data extracted. The data acquisition system sends out the sensor excitation signal and monitors the impedance changes found in a dielectric sensor. As shown in Figure 16-3, the sensor is representative of a flat plate capacitor. The capacitor area is fixed by the plate size, the capacitive distance is fixed by the filter paper between the two plates, and the capacitive medium changes from air to the resin system as the resin becomes viscous enough to wet the filter paper. The computer software, data acquisition system, and sensors were procured from MicroMet Instruments, Newton Centre, MA.

Two runs were made with the experimental set-up using a "neat" Bismaleimide V390 resin, available from U.S. Polymeric. "Neat" refers to a clean resin not applied to a structural carrier such as carbon fibers. Figures 16-4 and 16-5 show log e" information extracted from the sensor, plotted with the temperature versus time. Log e" is the loss

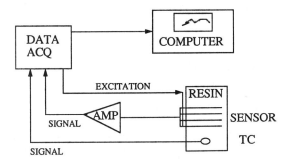

Figure 16-2. MicroMet Instrument Dielectrometer set-up.

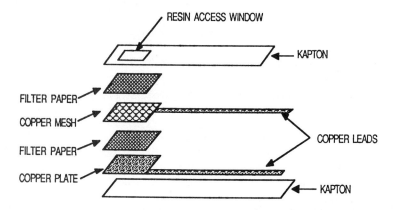

Figure 16-3. MicroMet Instruments Porous Parallel Plate Sensor.

factor associated with the plate capacitor. As the resin goes from a semi-hard to viscous to hard state, ionic mobility changes from zero to free to zero. When the free ions have some mobility, their relaxation times can be measured by introducing an excitation signal. This relaxation time can be used to calculate the loss factor which can be correlated somewhat to the resin viscosity and thereby to the amount of chemical reaction. See reference [2] for a full treatise. Data is taken throughout the cure at 1 Hz and 10 Hz. Note the magnitude difference in the response curves.

16.4   CONTROL SIMULATION

Once the data is stored from the process sensors it has to be manipulated into a form useful to the fuzzy logic simulator. A MicroMet supplied post process file translator puts the process data into an "ASCII" file. The ASCII file is then imported to a spreadsheet for the following manipulation.

Chap. 16 Fuzzy Logic Control of Resin Curing

- Extraneous data is deleted.
- The slope for log e" is calculated for both frequencies using linear regression over 5 points.
- Logical operator is used to provide a slope flag.
- All values are reduced to an integer with an appropriate bias.
- Supercalc four requires that a word processor be used to search and replace all commas with a space for a final data file that can be read by the fuzzy control system.

The Fuzzy Logic Controller (FLC) used for this project is the NeuraLogix Development System for the NLX-230. The FLC requires that all digital input data be integers from 0 to 255 and separated by spaces.

Configuration of the FLC was as follows.

- SYSTEM INPUTS; Time, Temperature, Hz, Slopeflag, Slope_value1, Slope_value2
- SYSTEM OUTPUTS; Immediate pressure signal at 100
- TERMS (or fuzzy memberships defined) All members inclusive (See Table 16-1).
- RULES MATRIX; The rules that infer output (See Table 16-2).

TERMS			Center	Width
Time	is	Tok	90	11
Temp	is	Tempok	143	17
Hzl	is	Hzlok	1	0
Slope_flag	is	Slopeflag	1	0
Slope_value1	is	Slope1	70	20
Slope_value2	is	Slope2	7	2

Table 16-1. Inputs fuzzy terms and specifications.

			Rule 1	Rule 2
Time	is	Tok	X	X
Temp	is	Tempok	X	X
Hzl	is	Hzlok	X	
Slope_flag	is	Slopeflag	X	X
Slope_value1	is	Slope1	X	
Slope_value2	is	Slope2		X

Table 16-2. NLX rules matrix.

Although two rules are simplistic, they gave the desired results. Rule one in effect states that if the material is in the process window of Time and Temperature and the slope is positive and within its required membership for the 1 Hz frequency then pressure can be applied. Rule 2 is of the same nature for the 10 Hz signal.

16.5 RESULTS

Figures 16-6 and 16-7 show the experimental data on an expanded scale with a designation of when a normal time dependent process would call for pressure and when the fuzzy logic simulation called for pressure. The first data set in Figure 16-6 shows the fuzzy logic controller calling for pressure at 87 minutes instead of the standard process time of 90 minutes. The second data set in Figure 16-7, responded to the simulation with a time of 90 minutes.

16.6 CONCLUSION

The resin and experimental setup was based on utilizing existing materials. Several more supporting runs should exercise the fuzzy rule implementation and modify or increase the number of rules based on further data. However, the experiment does indicate that fuzzy logic control could be used to determine a point of pressure application for a complex resin system. Note that use of the dielectric sensor is easier to implement in the laboratory with a neat resin than in a part with carbon particulates from the structural carrier. In the latter case, pressure must be used to force resin through a filter that restricts the carbon particulates that can short the sensor. This pressure is sometimes high enough to disturb the cure cycle that is under control. Advances in sensor technology will improve the reliability of any type of controller that uses chemical reaction monitoring input.

Currently application of sensors that monitor chemical reactions are normally used for incoming quality control. This quality check verifies that the material is within a processibility range. To make use of the supplied information already gathered, a fuzzy logic control simulator could optimize the material process by suggesting an optimized cure cycle for that batch of material.

Use of reaction monitor sensors during the actual cure cycle has become popular with the manufacturing of large costly composite parts in the aircraft industry. The same principles used in this report could enhance the quality of those parts by adding an in process controller based on fuzzy logic.

REFERENCES

[1] George R. Dvorsky, "The Practical Side of Dielectric Monitoring During Composite Cures," SME conference paper EM83-101 *Composites in Manufacturing II,* January, 1983.
[2] P. Ciriscioli, G. Springer, "Dielectric Cure Monitoring (A Critical Review)," SAMPE Journal, Vol. 25, No. 3, May/June 1989.
[3] MicroMet Instruments Inc. NEWSLETTER, Vol. 2, No. 2, Winter 1988.
[4] American NeuraLogix, Inc., Fuzzy MicroController data sheet for NLX-230, July 1991.

17

FUZZY LOGIC CONTROL IN FLIGHT CONTROL SYSTEMS

Craig Baker
Honeywell, Defense Avionics Systems Division

Aircraft pilots make decisions based on an enormous amount of information, but automatic flight control systems have limited amounts of information and time in which to make the same decisions. Yet, pilots expect flight control systems to perform up to the same standards as human pilots. Presently, flight control systems using classical control theory are unable to meet those high standards in some modes of operation. Classical control theory has inherent limitations which can not be overcome. Therefore, a totally different technique may be needed for a complete flight control solution. A promising solution to the flight control problem is fuzzy logic control. Fuzzy logic control uses information in the same manner as human experts, and fuzzy logic control does not require the complex mathematics associated with classical control theory. Unfortunately, the limitations and significance of fuzzy logic control are not widely understood. This paper explores the practical limitations and significance of fuzzy logic control as applied to flight control systems. The paper describes the design, implementation, and testing of a fuzzy logic controller for Very-high-frequency Omnidirectional Range (VOR) navigation in an automatic flight system for a transport aircraft. The design and implementation of the controller was successful, but the performance of the controller was not as robust as expected. The lack of robustness was due in part to the fact that the controller required more input information than was originally envisioned. Fuzzy logic control did improve the readability of the software, and it enhanced the autopilot's decision-making ability. In the future, fuzzy logic control could dramatically change the autopilot industry.

17.1 FUZZY LOGIC CONTROL IN FLIGHT CONTROL SYSTEMS

Honeywell's Defense Avionics Systems Division in Albuquerque, NM builds flight control systems, also known as autopilots, for a wide variety of aircraft. Honeywell uses classical control techniques for most systems, including a recent application for a large transport aircraft. This most recent autopilot system was extensively flight tested last summer with generally good results. One mode of operation which did not meet the performance criteria of the pilots was VOR navigation. Honeywell is currently working on a classical control theory solution to the VOR control problem, but due to the inherent limitations of classical control theory and the demands of the pilots, a more contemporary solution may be needed. Fuzzy logic control is a promising technique, and recent developments may make it a viable solution. Unfortunately, the limitations and significance of fuzzy logic control are not widely understood. This paper will explore the practical limitations and implications of fuzzy logic control as applied to flight control systems and will entail design decisions, fuzzy logic control implementations, and performance evaluations.

Due to the unique environment in which autopilots have to perform, some background information will be presented here to help the reader understand some of the design and implementation decisions which went into the fuzzy logic controller. VOR navigation is based on a fixed ground station which sends two radio signals. The phase difference between the signals determines an angle or radial to the VOR station. Figure 17-1 shows how the phase angle changes for the different radials around the station. The receiver on the aircraft determines the phase difference between the received signals, and compares it to a reference phase difference determined from the Course Select knob in the cockpit. The disparity between the reference phase difference and the received phase difference is annunciated on a display in the cockpit. The VOR signal's phase difference represents an angle, and it has no relationship to actual distance from the desired radial. A phase difference of 1 degree at 50 miles from the VOR station has 50 times more error than the same phase difference at 1 mile from the VOR station. Since the VOR signal can't differentiate distances, a separate signal is used to measure the distance from the VOR station. This signal is a basic radar system which is referred to as the Distance Measuring Equipment (DME). Military airplanes have the ability to use Tactical Aviation Navigation (TACAN). TACAN is equivalent to a VOR station which is always coupled to DME signals (VOR/DME). In this report, VOR will refer to VOR, VOR/DME and TACAN stations.

Chap. 17 Fuzzy Logic Control in Flight Control Systems

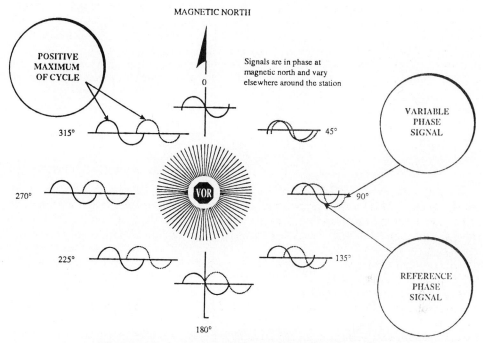

Figure 17-1. Signal Phase Angle Relationship.

Figure 17-2 shows a TACAN station combined with a VOR/DME station. The coupling of these two signals can not be taken for granted. Many VOR stations do not support DME signals, and both signals are not always operational when a station does support them. These inconsistent conditions help create the unique environment in which the fuzzy logic controller must work, and where fuzzy logic controllers can out perform classical controllers.

There are other factors which help create the unique environment in which the VOR fuzzy logic controller must work. These factors arise from the way in which pilots use VOR navigation. Pilots use VOR navigation for short and long range flights in all kinds of weather and pilots use VOR navigation for non-precision approaches into airports with VOR systems. Long range cruise conditions can have speeds over three times as fast as approach conditions, and the required settling time for cruise conditions is much less stringent than the settling requirements for approach conditions. At high altitude cruise conditions, it is not absolutely necessary to fly with no errors in course, or deviation from the radial. In fact, significant course errors may be necessary due to the weather conditions. Non-precision approaches are performed in airspace where the margin of error is very small, and in this airspace small errors can be fatal. In the airspace directly above or very close to a VOR station the VOR signal can be erratic and practically worthless for navigation. In this condition, unrealistic, rapid oscillations in VOR deviation are detected by the autopilot, and the mode of operation is called overstation.

366 Chap. 17 Fuzzy Logic Control in Flight Control Systems

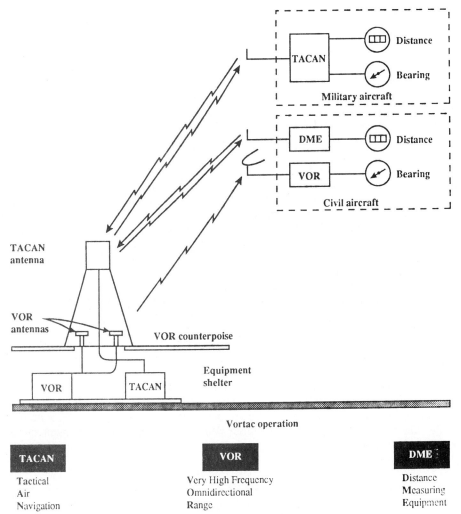

Figure 17-2. Vortac Operation.

When the operational mode is overstation, autopilots ignore the VOR deviation indications and they use their heading as a reference to fly from, until the VOR signal has settled. Once the VOR signal has settled, the autopilot will automatically return to using VOR deviation as a reference. Fuzzy logic could have a distinct advantage over classical control theory in detecting overstation conditions. Based on weather conditions, instinct and expertise, human pilots can determine when and to what extent errors are acceptable, but the autopilot must empirically determine acceptable errors. Determining acceptable errors is very fuzzy in nature, and it is another situation in which fuzzy logic controllers may out perform classical controllers. There are other specific elements to VOR navigation which make it a very difficult classical control problem, and these elements of

Chap. 17   Fuzzy Logic Control in Flight Control Systems

VOR navigation will be addressed along with the implementation of the fuzzy logic controller.

With varying weather conditions and flight conditions, VOR navigation is very dynamic. Classical controllers can be very effective when they are designed for a very rigid and specific purpose, but they may not perform well outside of those design boundaries. Expanding the boundaries of an established classical controller was the idea behind the first phase of this project, but as all things do, the idea has changed. The idea has grown into an independent fuzzy logic controller for the VOR problem. Originally, this project was going to be a fuzzy logic gain tuner for the VOR classical control solution which already exists in the autopilot. The fuzzy logic rules would dynamically adjust the gains in the autopilot to change the operational boundaries of the classical controller. As the design for the fuzzy tuner was being completed, a comparison of the requirements between a fuzzy logic tuner and a fuzzy logic controller showed the requirements of a fuzzy logic tuner closely resembled the requirements of a fuzzy logic controller. The required inputs were the same and the necessary outputs were similar in nature, but not in scale. The knowledge base was practically identical, but the fuzzy logic tuner was to be tremendously more complicated to practically debug and test. The fuzzy logic tuner required the knowledge of classical control theory and fuzzy logic. The fuzzy tuner would not be very maintainable due to the interactions between rules, gains and performance. The middle step appeared to be more trouble than it was worth, and the project was changed from a fuzzy logic gain tuner to a fuzzy logic controller.

In retrospect, the design requirements changed in the middle of the project. Originally, the project was only interested in improving the performance of the VOR control law. Now the project involves solving the VOR navigation problem in an efficient, understandable manner while improving performance. Several fuzzy logic controller designs were proposed, and a design which didn't require as many classical feedback loops was chosen for the initial development phase. Minimizing feedback loops reduces the required I/O, and maybe it will perform better. Granted, this idea sounds a bit crazy, but it did provide a good place to start building a fuzzy logic controller for VOR navigation. The fuzzy logic controller was designed with three inputs and one output. The inputs were VOR beam deviation, VOR beam deviation rate, and airspeed. The membership functions are shown in Figure 17-3. VOR deviation was needed to determine the direction and magnitude of the error. VOR beam deviation rate was necessary for overshoot suppression. Unrealistic beam rates were used to determine the overstation mode. Airspeed was used as an indicator of operational mode. The faster speeds indicated cruise modes and slow speeds represented approach modes. Figure 17-4 shows the membership functions for the output variable, VOR command, and Figure 17-5 shows the fuzzy associate memory diagram for VOR command. VOR command needed to determine size and direction of the necessary response. The knowledge base represents every possible combination of the membership functions. Unfortunately, the system was not stable. The design did not account for several important aircraft characteristics. Some of these characteristics were accounted for in the next fuzzy logic controller design.

Due to dynamic aircraft characteristics the design requirements for this project were changed, again. The design was now required to just match current autopilot

368 Chap. 17 Fuzzy Logic Control in Flight Control Systems

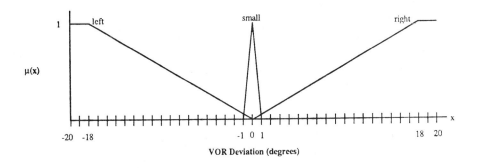

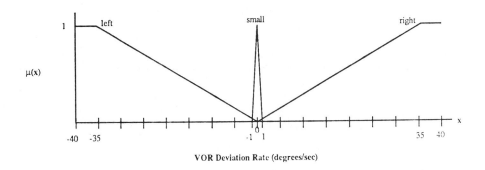

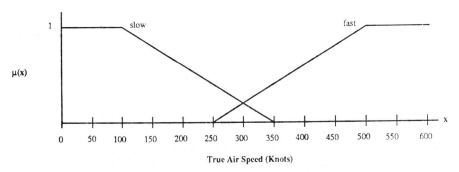

Figure 17-3. Input Membership Functions.

performance in cruise conditions, while utilizing a sophisticated but maintainable knowledge base. Several classical control feedback loops were replicated in the fuzzy logic controller, as well as a hierarchical command structure. The basic design remained the same, but more I/O was added to account for aircraft dynamics. The following inputs were added to the fuzzy logic control system. Cross track error was added to provide another measure of beam error. Cross track error represents the perpendicular distance the aircraft

Chap. 17 Fuzzy Logic Control in Flight Control Systems

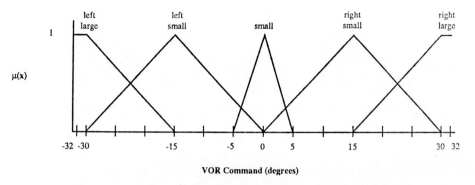

Figure 17-4. Membership Functions for VOR Command.

Figure 17-5. Fuzzy Associate Memory for VOR Command.

is from the desired course. Course error was added to provide feedback on which way the plane is pointed. The earlier fuzzy logic controller would close out the beam error, but the airplane would be pointed in the wrong direction, and it would overshoot the desired VOR radial very badly. The membership functions for these new inputs can be seen in Figure 17-6. The output changes were relatively minor. A second output was added, and the original output was renamed. The outputs were named VOR proportional command and VOR integral command. The membership functions for both outputs are identical to the membership function shown in Figure 17-4. The proportional command was designed to bring the aircraft close to the VOR beam, and the integral command was designed to gently close out the small errors. This requires another aircraft mode called VOR track. VOR track determined which VOR output command was appropriate for the conditions.

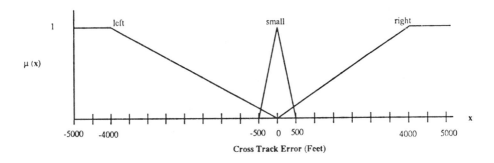

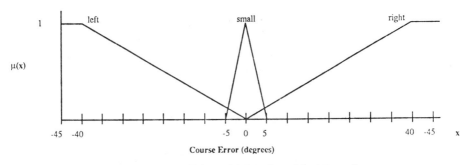

Figure 17-6. Additional Membership Functions.

When the aircraft had a small course error and a small deviation error, the aircraft was considered close enough to the beam for the VOR integral command. Otherwise, the VOR proportional command was used for navigation. The overstation logic remained the same as discussed earlier in the report. The fuzzy associate memory diagram became much more complicated with the addition of the new inputs. A section of the fuzzy associated memory diagram is shown in Figure 17-7. The entire diagram was considered too large for this report. As the figure shows, the knowledge base was shared by the VOR integral and VOR proportional commands. The difference between the two outputs was the input gains and the respective output gains. A gain of 0.5 was applied to VOR deviation rate for the proportional command to allow larger closing rates. A gain of 0.7 was given to course error to desensitize the fuzzy logic controller from course error at flight conditions away from the desired radial. The fuzzy logic controller worked the best with a proportional command gain of 0.8. For the integral command, the VOR rate and deviation gains were set to 2.0. All other input gains were set to 1.0. This was done to still be sensitive to deviation errors, but not to over control the aircraft. The fuzzy logic controller worked the best with an integral command gain of 0.6. The performance of the airplane was not as good as anticipated, but the maintainability of the software was tremendously improved.

The performance of fuzzy logic control was not as robust as anticipated, but this does not rule out possible improvements for the future. Figure 17-8 shows the time responses for the increase, and as the aircraft crossed the desired radial the aircraft was not

Chap. 17 Fuzzy Logic Control in Flight Control Systems

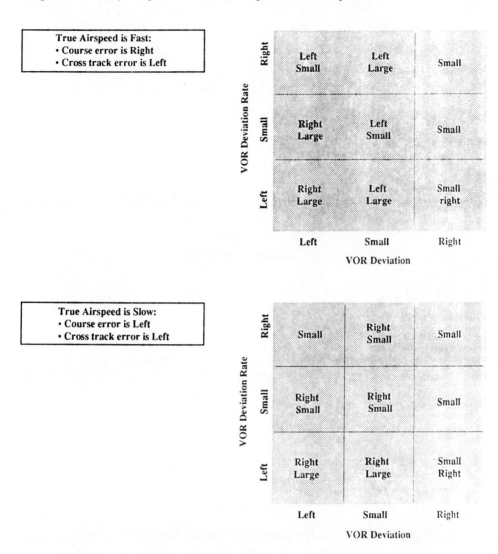

Figure 17-7. Fuzzy Associate Memory for Output Command.

pointed in the direction of the desired radial. This caused large a overshoot and ultimately, the fuzzy logic controller was unstable. The second design anticipated course error problems too well. As the plot shows, the response is over damped, and it takes too long for the aircraft to settle, even in long cruise conditions. The implementation of the two fuzzy logic controllers. In both cases, the aircraft was initialized with a five degree course into the desired VOR radial at 30 miles from the station. The aircraft had a constant velocity of 475 knots and altitude was held at 30,000 feet. The plot for the initial design shows the ill effects of not correcting for course errors. As the deviation decreased, the course error performance of the operational mode logic was very good. The fuzzy logic

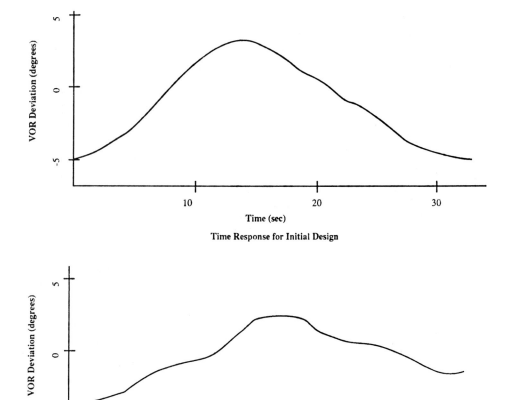

Figure 17-8. Time Responses for Fuzzy Logic Controllers.

rules used to determine modes were very simple, but they performed well. If more information was added to the control and mode logic rules, the performance of the VOR navigation fuzzy logic controller would only get better. Even though the performance of the fuzzy logic controller did not meet the anticipated results, the project was very valuable. The project did show some insight into fuzzy logic control, and the project highlighted some of the benefits of fuzzy logic control. Fuzzy logic control is not a replacement for classical control techniques. As this project illustrates, fuzzy logic control requires the same kinds of feedback loops, and gains necessary in classical control theory. Fuzzy logic control provides the same kind of gain scheduling found in classical control theory, but fuzzy logic control is implemented in a less precise, easily understandable manner. The project demonstrated how fuzzy logic controllers are easy to debug, and easy to design. Also, this project demonstrated how easily operational modes can be determined with fuzzy logic. These qualities are very important when the autopilot is being tested or modified long after the initial design phases of the program. This project only investigated one small part of the lateral axis for the autopilot system. There are numerous other places where fuzzy logic and fuzzy logic control could be applied to autopilots. Pilots are continuously making fuzzy decisions with regard to control and safety. There is no reason to believe that fuzzy logic could not handle those same situations if the same information was available to the autopilot system. In the future, all of the autopilot control laws and mode logic could be implemented in fuzzy logic, and possibly one generic autopilot could control several different kinds of aircraft.

REFERENCES

Federal Aviation Administration. *Instrument Flying Handbook*. U.S. Department of Transportation, 1980.

Knuth, D.E. *Seminumerical Algorithms, The Art of Computer Programming*. Addison-Wesley, 1981.

Larkin, L.I. "A Fuzzy Logic Controller for Aircraft Flight Control." Proceedings of 23rd Conference on Decision and Control, December, 1984, pp. 894 -897.

Rolfe, J.M. and K.J. Staples. *Flight Simulation*. Cambridge University Press, 1988.

18

TUNING OF FUZZY LOGIC CONTROLLERS BY PARAMETER ESTIMATION METHOD

Nahrul K. Alang Rashid and A. Sharif Heger
University of New Mexico

Fuzzy logic controllers (FLC) require fine tuning to match the rules to the membership functions or vice-versa. For the class of FLCs that mimic human process operators, the rule-membership function mismatch arises from the lack of information on the specifications of the membership functions. The rules that are incorporated into the FLC knowledge base are broad generalizations of the operators' control strategy. While the rules are readily available from the operator, the specifications of the membership functions are harder to define. For the class of FLCs that are used to control a process in which the control actions are not known a-priori, the rules and membership functions are derived using heuristics or based on the dynamics of the process that are obtained using simulation models. In this class of FLCs, and the one mentioned earlier, overlaps between variables fuzzy subsets, the slopes, and the functions used in defining the membership values all tend to dilute the generality of the rules and introduce specifics to the FLC.

Chap. 18 Fuzzy Logic Controllers by Parameter Estimation

This paper presents a method for tuning FLCs that mimic the actions of human process operators. The tuning method treats the FLC as a dynamic system by itself, transformed from the dynamics of the process to be controlled, with some parameters in which their optimal values are to be determined. Estimation of these parameters constitutes the tuning procedure. The estimation is conducted by using multiparameter least square estimation method. In order to make this paper self-contained a brief review of FLC is first presented. The tuning method is then described and example of its application is illustrated by using a process simulation model.

18.1 INTRODUCTION

FLCs are characterized both by their structures and types. Basically, the structure is identified by the fuzzification and the defuzzification methods used. Some fuzzification and defuzzification schemes that are commonly used have been elaborated elsewhere, see for example Lee [1,2] or Kosko [3].

The types of FLCs are classified by the way in which they are used. Following Zadeh [4], FLCs that mimic human process operators are called Type-1. The other type of FLCs, which we will call Type-2, are those that perform control tasks that are not normally performed by human process operators. Figures 18-1a and 18-1b highlight the difference between these two types of FLCs. Notice that in Type-2 FLCs the desired controller output is not known whereas in Type-1 FLCs the controller output is known, that is that of human process operators. Thus, the purpose of tuning in the former is to let the controller adjust its outputs such that the plant response approaches the desired values. In Type-1 FLCs, on the other hand, the purpose of tuning is to make the controller outputs follow human operators control actions. In this paper, we present a method for tuning FLCs of Type-1.

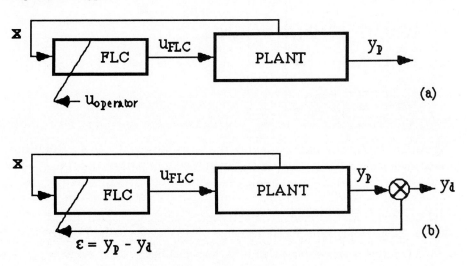

Figure 18-1: Two types of FLC:
(a) Type 1 - tuning is for adjusting $u_{FLC} = u_{operator}$.

(b) Type 2 - tuning is for finding u_{FLC} such that $y_p = y_d$.

The process of designing FLCs begins with the extraction of typical control rules. For the Type-1 FLCs these rules are obtained from human process operators. The rules are typical if they are representative of the control strategy used by the process operators who are responsible for operating the process. The membership functions for each of the fuzzy subsets appearing in the rules are then specified. The performance of the FLC is closely influenced by the rules and by the closeness of the membership functions to the true reflections of the statement of the rules. While the rules are known, the operator may find it difficult to specify the meaning of the linguistic values that he uses.

Notations

The FLC structure that we will use in this paper is shown in detail in Figure 18-2. For simplicity and clarity, the FLC being considered has only two inputs and one output.

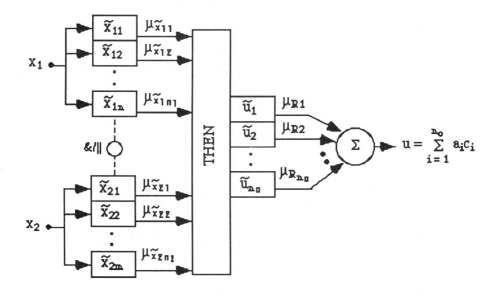

Figure 18-2: A schematic showing the workings of the FLC. Relations between the inputs fuzzy subsets in the antecedent part of the rule is shown by the &/|| sign.

To set our notations, continuous variables are denoted by capital letters, e.g., X; numerical values of the variables in the range where they are defined by small letters, e.g., $x \in [x_L, x_U]$ is a possible value of X that is defined in the range of $[x_L, x_U]$; linguistic variables by $\widetilde{X}$; and its fuzzy subsets, or linguistic values by $\widetilde{x}_j$, $j = 1, n$, where n is the number of fuzzy subsets in $\widetilde{X}$, thus $\widetilde{X} = \widetilde{x}_j$.

Chap. 18 Fuzzy Logic Controllers by Parameter Estimation

An mth rule is denoted by R_m, and stated in the form of fuzzy conditional statement shown in equation (1). In this equation, X_1 and X_2 are the input variables and U is the output variable.

$$R_m : \text{if } X_1 \text{ is } \tilde{x}_{1j} \text{ and } X_2 \text{ is } \tilde{x}_{2j} \text{ then } U \text{ is } \tilde{u}_j \qquad (1)$$

Equation (1) may correspond to: "if temperature is high and flow_rate is low then pump_speed is high," which means increase the pump speed if temperature is high and flow rate is low. Thus, in our notation; X_1 = temperature; X_2 = flow_rate; and U = pump_speed. The corresponding linguistic variables and values may, for example, be; $\tilde{X}_1$ = $\{\tilde{x}_{11}, \tilde{x}_{12}, \tilde{x}_{13}\}$ = {low,medium,high}; $\tilde{X}_2$ = $\{\tilde{x}_{21}, \tilde{x}_{22}\}$ = {low,high}; and $\tilde{U}$ = $\{\tilde{u}_1, \tilde{u}_2, \tilde{u}_3\}$ = {decrease,no_change,increase}.

This rule can be used in a FLC only when the membership functions, $m_{\tilde{x}_{ij}}$ and $m_{\tilde{u}_j}$ of each variables fuzzy subsets are specified. Then, any $X_i = x_i$ can be said to be $\tilde{x}_{ij}$ to the degree $m_{\tilde{x}_{ij}}(x_i)$. Two of the most widely used membership functions are the triangle and the Gaussian functions. In this work, the Gaussian membership function is used. The slopes of each fuzzy subset's membership functions represent the degree to which variable values within that fuzzy subset belongs to that particular subset. The overlaps between fuzzy subsets membership functions represent the degree to which the variables within that overlapping regions belong to one or the other subsets. The specification of these parameters is part of the FLC design process, and will therefore affect the performance of the FLC.

In Figure 18-2, the degree to which the input variables $X_i = x_i$, i = 1, 2 belong in their respective linguistic values are first determined by the corresponding membership functions, $m_{\tilde{x}_{ij}}$, j = 1, n_i, where n_i is the number of fuzzy subsets in the variable i. The degree to which each rules R_m, m = 1, n_R, where n_R is the number of rules, is activated by this input pair is determined by either equation (2a) or equation (2b).

$$\mu_{R_m}^{(\tilde{u}_j)}(x_1, x_2) = \vee (\mu_{\tilde{x}_{1j}}(x_1), \mu_{\tilde{x}_{2j}}(x_2)) \qquad (a)$$
$$= \wedge (\mu_{\tilde{x}_{1j}}(x_1), \mu_{\tilde{x}_{2j}}(x_2)) \qquad (b) \qquad (2)$$

Equation (2a) is used when the antecedent of the rule shown in equation (1) contains an "OR" connective and equation (2b) is used for the "AND" connective. The superscript indicates the output fuzzy subset to which this rule belongs to.

The FLC output is calculated by using a simplified centroid defuzzification scheme shown in equation (3),

$$u = \sum_{j=1}^{no} a_j c_j \qquad (3)$$

where n_o is the number of the output fuzzy subsets and c_j is the centroids of the output fuzzy subsets membership function. Equation (3) shows that the FLC output is an algebraic equation made up of the parameter c_j that is scaled by the weighted average of the rule activation strength. This scheme is not the only defuzzification method available, but is one of the most widely used because of its simplicity.

The centroids of the output variable fuzzy subsets membership functions are given by equation (4).

$$\int_{u=u_L}^{u=c_j} m_{\tilde{u}_j} \, du = \int_{u=c_j}^{u=u_H} m_{\tilde{u}_j} \, du \qquad (4)$$

The parameter a_j measures the degree to which each of the output fuzzy subsets contributes to the FLC output. The values of a_j are calculated by using equation (5).

$$a_j = \frac{m_{R_j}}{\sum_{j=1}^{n_o} m_{R_j}} \qquad (5)$$

Since the number of rules, n_R can be more than the number of output fuzzy subsets, n_o equation (2) cannot be used directly in equation (5). The selection of which $m_{R_m}^{(\tilde{u}_j)}(x_1, x_2)$ is to be used for each $\tilde{u}_j$, $j = 1, n_o$ is done by taking the maximum of the rule activation strengths according to equation (6).

$$m_{R_j} = \left(m_{R_m}^{\tilde{u}_j} \right), \text{ " } m = 1, n_R \qquad (6)$$

18.2   FLC TUNING

FLCs can be tuned by modifying the rules, the membership functions, or both rules and membership functions. Rule modifications, however, may give rise to rules that are not consistent with those that are used by the operator. Consequently, maintenance, troubleshooting, or modifications of the FLCs at a later date may become time consuming and troublesome.

In this section we introduce a method for membership function modification which is based on parameter estimation technique. The basic idea of this method is by first treating the FLC as a dynamic system with some parameters that need to be optimized. The mathematical expression for this system is described by the FLC defuzzification equation, which in fact is its input-output relationship in the numerical domain as shown in equation (3). This equation is the process-FLC closed loop system dynamic since it is a transformation of the control rules from linguistic to algebraic forms. The control rules, inherently, embody the dynamic of the process to be controlled and the desired control action.

Chap. 18 Fuzzy Logic Controllers by Parameter Estimation

In Type-1 FLCs the desired controller outputs are known for given plant states. Let the known controller outputs, as given by the plant operator, for plant states $\mathbf{x}_k$ be u_{dk}, $k = 1, n_s$ where n_s is the number of time steps. The underlying assumption is that by applying u_{dk} the desired plant response, y_{dk} is obtained. Due to the rule-membership function mismatch enumerated earlier, the untuned FLC output, u_k however, is not necessarily equal to u_{dk}. Let the error between the desired controller output and the untuned FLC output be e_k. Then, the untuned FLC output can be written as in equation (7).

$$\begin{aligned} u_1 &= a_{11}c_1 + a_{21}c_2 + \cdots + a_{n_o 1}c_{n_o} + e_1 \\ u_2 &= a_{12}c_1 + a_{22}c_2 + \cdots + a_{n_o 2}c_{n_o} + e_2 \\ &\vdots \\ u_{n_s} &= a_{1n_s}c_1 + a_{2n_s}c_2 + \cdots + a_{n_o n_s}c_{n_o} + e_{n_s} \end{aligned} \quad (7)$$

where a_{ik}, $i = 1, n_o$ and $k = 1, n_s$ is given by equation (5), and c_i, $i = 1, n_o$ is determined by equation (4). Using vector notation, the untuned FLC output at any time step in equation (7) can be written as in equation (8).

$$u_k = \mathbf{a}_k^T \mathbf{c} + e_k \quad (8)$$

The vectors $\mathbf{a}_k$ and $\mathbf{c}$ are given by equations (9) and (10), respectively.

$$\mathbf{a}_k^T = [a_1 \; a_2 \; \cdots \; a_{n_o}]_k \quad (9)$$

$$\mathbf{c}^T = [c_1 \; c_2 \; \cdots \; c_{n_o}] \quad (10)$$

The objective of tuning the FLC is to minimize the sum of the square of the errors in equation (7) by modifying the centroid vector, $\mathbf{c}$. Let the modified centroid vector at the end of the tuning procedure, $\mathbf{c}^*$ be given by equation (11).

$$\mathbf{c}^{*T} = [c_1^* \; c_2^* \; \cdots \; c_{n_o}^*] \quad (11)$$

Then, the new FLC output shown in equation (12) minimizes the objective function shown in equation (13).

$$u_k = \mathbf{a}_k^T \mathbf{c}^* \quad (12)$$

$$J = \sum_{k=1}^{n_s} \left[\mathbf{a}_k^T \mathbf{c}^* - u_{dk} \right]^2 \quad (13)$$

By letting $\partial J / \partial \mathbf{c}^* = 0$, the normal equation for the multiparameter least square estimation problem is given by:

$$\left[\sum_{k=1}^{n_s} \mathbf{a}_k \mathbf{a}_k^T\right] \mathbf{c}^* - \sum_{k=1}^{n_s} \mathbf{a}_k u_{dk} = 0. \quad (14)$$

Solving for the optimum centroid vector, $\mathbf{c}^*$ we obtain:

$$\mathbf{c}^* = \left[\sum_{k=1}^{n_s} \mathbf{a}_k \mathbf{a}_k^T\right]^{-1} \sum_{k=1}^{n_s} \mathbf{a}_k u_{dk}$$
$$= \mathbf{R}_{n_s} \mathbf{p}_{n_s} \quad (15)$$

where $\mathbf{R}_{n_s}$ is an ($n_s \times n_s$) matrix given by equation (16) and $\mathbf{p}_{n_s}$ is a ($n_s \times 1$) vector given by equation (17).

$$\mathbf{R}_{n_s} = \left[\sum_{k=1}^{n_s} \mathbf{a}_k \mathbf{a}_k^T\right]^{-1} \quad (16)$$

$$\mathbf{p}_{n_s} = \sum_{k=1}^{n_s} \mathbf{a}_k u_{dk} \quad (17)$$

Equation (15) gives the centroid vector for the tuned FLC obtained for a collection of n_s observations. This equation can be written in recursive form by first writing equations (16) and (17) recursively as in equations (18) and (19), respectively.

$$\mathbf{R}_{n_s}^{-1} = \mathbf{R}_{n_s-1}^{-1} + \mathbf{a}_{n_s} \mathbf{a}_{n_s}^T \quad (18)$$

$$\mathbf{p}_{n_s} = \mathbf{p}_{n_s-1} + \mathbf{a}_{n_s} u_{dn_s} \quad (19)$$

Through several algebraic manipulations of equations (18) and (19) the recursive form of the optimum centroid vector that is shown in equation (20) can then be derived. For brevity, these manipulations are not shown in this paper as the details have been elaborated by Young [5].

$$\mathbf{c}^*_{n_s} = \mathbf{c}^*_{n_s-1} - \mathbf{R}_{n_s} \mathbf{a}_{n_s} \left[\mathbf{a}_{n_s}^T \mathbf{c}^*_{n_s-1} - u_{dn_s}\right] \quad (20)$$

This equation shows that the FLC adapts its centroid vectors to the plant dynamics. The centroid vectors are updated as new plant states are available while keeping the centroid vectors optimum in the sense of the objective function shown in equation (13). It also needs a starting value which can be set to correspond to the initially designed value of the centroid vectors.

18.3 SIMULATION

To illustrate the methodology described above we will use a process simulation model. The process is a fluid holding tank. The control objective is to track or maintain the fluid

Chap. 18 Fuzzy Logic Controllers by Parameter Estimation

level, h so that it is close to a desired level, h_d. The set up of this process, adapted from Slotine and Li [6] is shown in Figure 18-3.

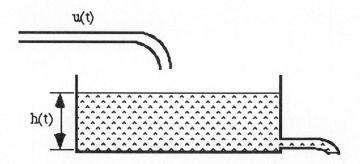

Figure 18-3: Fluid level control. Flow rate into the tank, u(t) is the manipulated variable, and the fluid level, h(t) is the controlled variable.

The dynamic of the system is given by equation (21), where g = 9.81 m/s², A(h) is the cross-sectional area of the tank, a is the cross-sectional area of the outlet pipe, and u(t) is the mass flow rate into the tank. The mass flow rate out of the tank is assumed constant.

$$A(h)\frac{dh}{dt} = u(t) - a\sqrt{2gh(t)} \qquad (21)$$

Derivation of FLC Rules and Membership Functions

Since we do not have an operator from which the rules and the membership functions for the FLC can be derived, we generate the desired controller output by using feedback linearization. Following Slotine and Li [6], the controller output for the process, u(t) is chosen as in equation (22). The parameter ∂ is a strictly positive constant. Substitution of equation (22) into equation (21) yields the closed-loop system dynamic, equation (23), which is used as the reference system in this example.

$$u(t) = a\sqrt{2gh(t)} - \partial A(h)(h(t) - h_d) \qquad (22)$$

$$\frac{dh}{dt} + \partial (h(t) - h_d) = 0. \qquad (23)$$

The closed loop system is simulated by using fourth-order Runge-Kutta algorithm. With Figure 18-4. The controller output, u(t) is calculated using equation (22). Both the process response and the controller output for this particular operation will be used to guide us in designing a basic FLC that will be tuned later for this process. In particular it provides information on the range of the controller output and the responsiveness of the process.

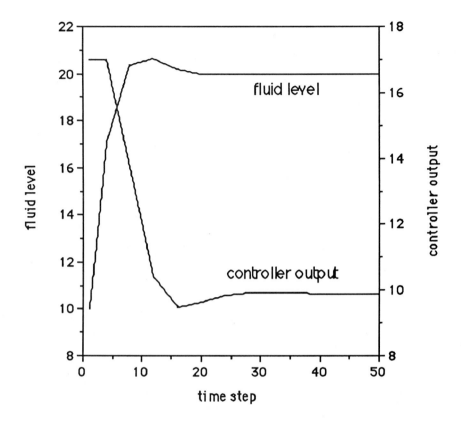

Figure 18-4: The process and the linearized feedback controller responses to a step demand in fluid level from 10 to 20 height units.

Untuned FLC, FLC-1

Based on the plant simulation model in Section 18-3 we design the FLC for the same control objective. The FLC has as its inputs the ERROR between the actual and desired fluid levels, and mass flow rate, FLOW as its output. The ERROR is defined by equation (24). It is therefore a single-input single-output FLC. In actual control of the process, human operator may also use the rate of error change as an additional input.

$$\text{ERROR} = h(t) - h_d \qquad (24)$$

By setting the tank height at 30 units, we define the ERROR range to be [-30,30]. From the process model we set the controller output, FLOW to be in the range of [-8,10]. Further, we assign five fuzzy subsets to each FLC variables. Let these fuzzy subsets be; NB: Negative_Big; NS: Negative_Small; ZE: Zero; PS: Positive_Small; and PB: Positive_Big. The membership function is "Gaussian-like" as given in equation (25).

$$m_{\tilde{x}}(x) = \left(\frac{-(x-\bar{x})^2}{2s}\right) \quad (25)$$

The parameters of the membership functions, the mean and the spread, are given in Table 18-1. The FLC rule base is shown in Table 18-2.

Equation (21) is used to simulate the system using the same integration algorithm. The controller output, or the manipulated variable, however, is now given by the FLC. Thus,

$$A(h)\frac{dh}{dt} = \tilde{f}(R,m) - a\sqrt{2gh(t)} \quad (26)$$

The function $\tilde{f}(.)$ is the FLC mapping function operating on its rule base, R and variable membership functions, m. The rule base is a function of the FLC input and output variables, $R = g(\text{ERROR,FLOW})$ shown in Table 18-2. We assume that this is the first cut of the FLC design which is based on the information on the rules and membership functions obtained from our "fictitious" human operator, which is the linearized feedback controller described in Section 18-3.

FUZZY SUBSETS	ERROR	FLOW
NB	-30 (200)	-8 (10)
NS	-15 (200)	-5 (10)
ZE	0 (200)	0 (10)
PS	15 (200)	5 (10)
PB	30 (200)	10 (10)

Table 18-1. Membership function parameters of the fuzzy subsets showing the mean, $\bar{x}$ and the spread, σ in parentheses.

ERROR	NB	NS	ZE	PS	PB
FLOW	PB	PS	ZE	NS	NB

Table 18-2. FLC Rule Base. The first column reads: if ERROR is NB then FLOW is PB.

The response of the process to this first-cut controller, FLC-1 for the same control objective as for the linearized feedback controller is shown in Figure 18-5. It is characterized by a damped oscillation with an initially large overshoot of about 40% of the desired fluid level. The FLC is designed and simulated using a fuzzy logic code program that is described in Alang-Rashid and Heger [7]

Ability of the FLC to track a time-varying desired fluid level is shown in Figure 18-6. The desired fluid level, shown in Figure 18-7 is varied according to $h_d(t) = 10 + 10 * (1$

- exp(- t/20)). Also plotted in Figure 18-6 is the FLC output, u_{FLC-1}. The FLC is designed so that its output is integrated. This explains the difference in the general trends of the FLC output and the linearized feedback controller output shown in Figure 18-4.

Figure 18-7 shows that the FLC is able to meet this control objective. The fluid level, however, settles at a slightly higher value than that prescribed by $h_d(t)$. Nevertheless, this shows that the rules and the membership functions are general enough for control objective that is different from the one that the FLC were designed for originally.

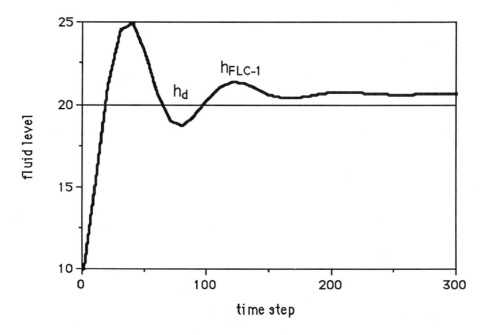

Figure 18-5: The process response to a unit step demand in fluid level from 10 to 20 height units by using FLC-1.

Tuned FLC, FLC-2

That the performance of the untuned controller, FLC-1 is close to the desired one suggests that it can be fine-tuned using the parameter estimation method. Application of this tuning method to FLC-1 shifts the centroid vector from $c = [\ -7.6\ -5\ 0.2\ 5.2\ 9\ ]^T$ to $c^* = [\ -7.6\ -3.75\ 0.08\ 3.4\ 9\ ]^T$. The tuning uses the linearized feedback controller output in tracking the $h_d(t)$ trajectory given earlier as the desired controller response. It is seen that the values of c_1 and c_5 do not change. This implies that over this operating region, the contributions of the output fuzzy subsets $\tilde{u}_1$ and $\tilde{u}_5$ to the controller output, u are negligible.

Chap. 18 Fuzzy Logic Controllers by Parameter Estimation

The response of the tuned FLC, which we denote as FLC-2, is shown in Figure 18-8. Comparison with the desired fluid level, Figure 18-9 shows that the offset error that appears in FLC-1 is eliminated.

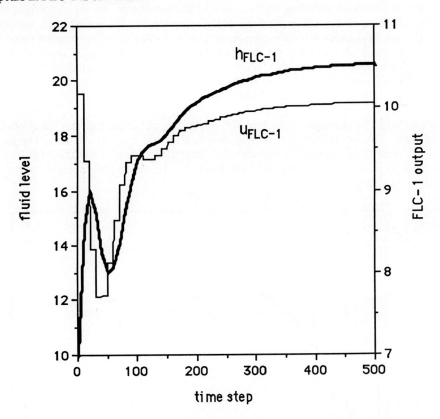

Figure 18-6: The process and FLC-1 responses in tracking a time-varying desired fluid level. The demanded fluid level is shown in Figure 18-7.

Other Desired Fluid Level Trajectory

From figures (6) through to (8) we can conclude that both the untuned, FLC-1 and the tuned, FLC-2 controllers perform as expected. The initial overshoot and undershoot in the controlled fluid level that appear in these figures is due to the disturbance in the flow rate at the onset of the FLCs initiation. In other words, the fluid level was not at a steady state condition when the FLCs are turned on since the control action, u was higher than that required to maintain the fluid level at 10 height units. The action of the FLCs to bring this down can be seen in both figures (7) and (8). By setting the correct initial condition, $u(t = 0^-) = a\sqrt{2gh(t)}$, the FLC performs smoothly as shown in Figure 18-10.

To extend the simulations further, first the desired fluid level trajectory is modified to include an increase and a decrease in the desired level. The increasing part is as given earlier and the decreasing part is described by -5 * (1 - exp((300 - i)/30)). This change is applied at time step, i of 300. As shown in Figure 18-11, the tuned FLC, FLC-2 performs much better than its untuned counterpart, FLC-1. The delay in the FLCs responses is due to the sampling interval used in the simulation. In any sampled-control system, sampling interval is one of the design parameters. In FLC design, sampling interval is a consideration for setting the membership functions of the controller variables. This is especially so when the variable explicitly contains time factor such as rate of change. Further, as evidenced by Figure 18-11, sampling interval governs the state of readiness of the controller. All simulations in this work have the same uniform sampling interval of 10 time steps with a Δt between time steps of 0.2 second.

Next we simulate an increase in the rate of fluid flowing out of the tank instead of holding it constant. This is equivalent to a reduction in the fluid level in the tank. The tuned controller, FLC-2 is required to maintain the same desired fluid level that was used in Figure 18-10. The result of this simulation, Figure 18-12, shows that the tuned controller performs satisfactorily.

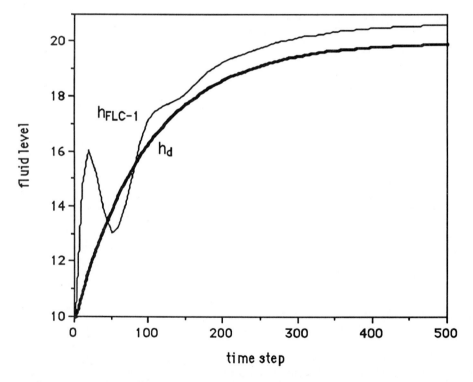

Figure 18-7: Comparison of the desired fluid level trajectory with that obtained by using the untuned controller, FLC-1.

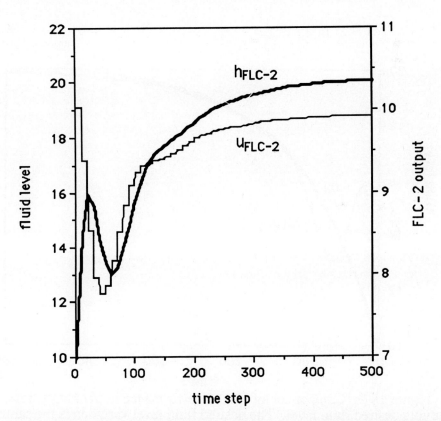

Figure 18-8: The process and controller responses in tracking a time varying desired fluid level. The controller is FLC-2, which is fuzzy controller FLC-1 that has been tuned. The desired fluid level is shown in Figure 18-9.

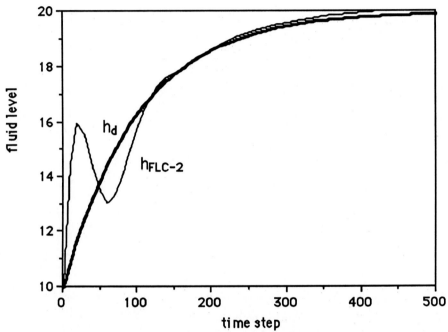

Figure 18-9: Comparison of FLC-2 performance in tracking a time-varying desired fluid level. The desired fluid level variation is the same as the one used for FLC-1 shown in Figure 18-7.

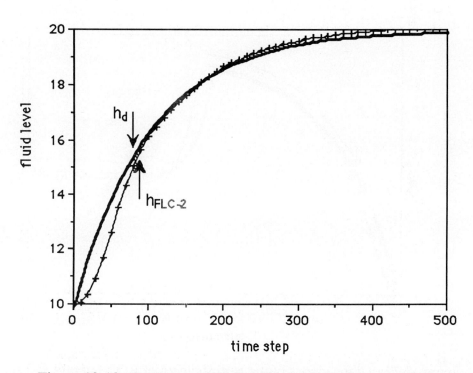

Figure 18-10: Response of FLC-2 after initial value correction.

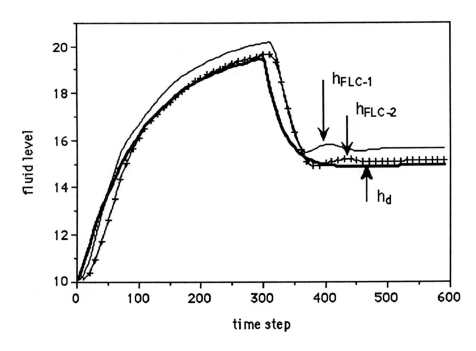

Figure 18-11: Comparisons of the performances of FLC-1 and FLC-2 in tracking the increase and decrease in desired fluid level.

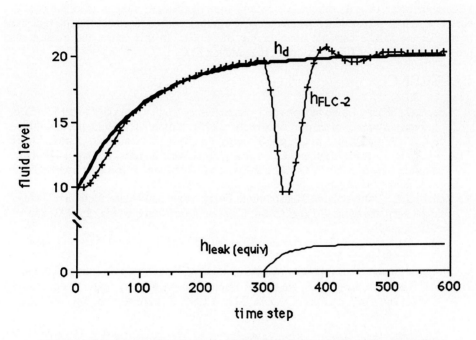

Figure 18-12: Performance of FLC-2 following an increase in fluid outflow.

18.4 CONCLUSIONS

A method for tuning Type-1 FLCs was presented. In summary, this method treats the FLC as a dynamic system in which its input-output relationship, that is expressed by its defuzzification scheme, contains parameters that can be optimized. The objective is to minimize a cost function which is the square of the sum of the errors between the controllers output and the desired output. On this premise, a regressive multiparameter least square parameter estimation method is used to set the unknown parameters.

Results of the simulations show that this tuning method improves the performance of the FLC while preserving its rule base. Optimization of the FLC's output variable centroid vector, which is the estimated parameters used in this method, does not require redesigns of any of the FLC's variable membership functions. In this sense, the method also preserves the membership functions of the untuned FLC.

REFERENCES

1. Lee, C.C., "Fuzzy Logic in Control Systems: Fuzzy Logic Controller - Part I," *IEEE Trans. Systems, Man, and Cybernetics*, 20, 2, March/April 1990, pp. 404-418.
2. Lee, C.C., "Fuzzy Logic in Control Systems: Fuzzy Logic Controller - Part II," *IEEE Trans. Systems, Man, and Cybernetics*, 20, 2, March/April 1990, pp. 419-435.
3. Kosko, B., *Neural Networks and Fuzzy Systems: A Dynamical Systems Approach to Machine Intelligence*, New Jersey: Prentice Hall, 1992.
4. Zadeh, L.A., "Interpolative Reasoning in Fuzzy Logic and Neural Network Theory," plenary talk at the *IEEE First Int. Conf. on Fuzzy Systems*, March 1992, San Diego, CA.
5. Young, P., *Recursive Estimation and Time Series Analysis: An Introduction*, New York: Springer-Verlag, 1984.
6. Slotine, J.E. and W. Li, *Applied Nonlinear Control*, New Jersey: Prentice Hall, 1991.
7. Alang-Rashid, N.K. and A.S. Heger, "A General Purpose Fuzzy Logic Code," *IEEE First Int. Conf. on Fuzzy Systems*, March 1992, San Diego, CA., pp. 263-268.

SUBJECT INDEX

A

AC motor, 329
Alignment, fuzzy control of laser beam, 149
Antecedents, 217
Applications of fuzzy logic, 1, 112
   autopilot, 363
   conveyor belt in robotics, 112
   detection of personnel, 217
   digital filtering, 232
   inverted pendulum, 112
   laser beam tracking, 149
   manufacturing planning, 171
   oil recovery, 181
   pattern recognition, 279
   power generation, 329
   resin curing, 357
   robot manipulators, 292
   thermal systems, 112
   traffic control, 262
   tuning by parameter estimation, 374
Approximate reasoning, 36
Autopilot control, 363

B

Bismaleimide, 357

C

C++ programming, 262
Cardinality
   crisp relations, 10
Calculus, Propositional, 10
Centroid defuzzification, 1, 36, 51, 86, 217
Crisp sets, 10, 217
Comparison
   flight, 363
   fuzzy and crisp control, 149
   fuzzy and crisp logic in oil recovery, 181
   traffic control, 262
Control
   fuzzy, 1, 51, 86, 112, 149, 262, 292, 329, 363
   traffic, 262
Conveyor belt, 112
Cross link, 357

D

DC motor, 329
Defuzzification methods, 1, 51, 86, 112, 374
Detection of personnel, 217
Dielectric sensors, 357
Diagram, Saggital, 10
Digital filtering, use of fuzzy logic, 232

E

Electric power generation, 329
Expert systems
   comparison with fuzzy, 181
   in oil recovery problem, 181
Extension Principle, 10

F

FAM - Fuzzy associate memory, 51, 112, 149, 363
FIPS - fuzzy inference per second, 112, 217
FLCG fuzzy software, 112, 374
Flight control, 363
FULDEK fuzzy software, 112
Fuzzification, 36, 51, 86, 112, 374
Fuzzy adaptive rules, 1, 51, 86

software, 112, 149, 262, 374
hardware, 112, 149, 329
  NeuraLogix microcontroller, 112, 357
  Togai single-board controller, 112
controller, 1, 51, 86
  conveyor belt, 112
  laser beam, 149
  NeuraLogix microcontroller, 112, 357
  robot manipulator, 292
  thermal systems, 112
  tuning, 374
controller design, 51, 86
expert systems, 51, 86
inferencing, 1, 36, 51, 86
logic, 36
  basic robot controller, 1, 51, 86
  comparison with crisp in oil recovery, 181
  control, 1, 51, 86, 112, 149
  digital filtering, 232
  hardware, 112, 149
  manufacturing planning use of, 171
  oil recovery use of, 181
  robot manipulators, 292
  software, 112
  tautologies, 36
  traffic flow control of, 262
  tuning by parameter estimation, 374
membership functions, 10
sets, 10
  convex, 10
  manufacturing planning use of, 171
  operations, 10
  properties, 10

G

Genetic algorithms, 86, 279
Geometric shape recognition, 279

H

Hardware of fuzzy logic, 112, 149, 292
  NeuraLogix microcontroller, 112, 357
  Togai single-board fuzzy controller, 112

I

Introduction
  fuzzy control, 1
  fuzzy logic, 1
  fuzzy sets, 1
Interpolation between control surfaces, 1
Interpolative reasoning, 112
Inverted pendulum, 112
Ions, 357

J

K

L

Label, linguistic, 217
Learning in fuzzy pattern recognition, 279
Logic, crisp, 36
Logic, fuzzy, 36

M

MATLAB, 112
Manufacturing planning use of fuzzy logic, 171
Matrix, fuzzy rules of, 51, 86, 149, 357
Membership function, 10
MicroMet, 357

N

NeuraLogix microcontroller, 112, 357

O

Oil recovery, enhanced, 181

Index

P

Parameter estimation, use in fuzzy control, 374
Pattern recognition, 279
Personnel detection use of fuzzy logic, 217
Planning in manufacturing, 171
Polymetric, U.S., 357
Power generation control, 329
Predicate logic, 36
Propositional calculus, 36

Q

R

Reasoning, approximate, 36
Recovery, oil using fuzzy logic, 181
Resin curing, 357
Rhino robot, 292
Robot manipulator, fuzzy control of, 292
Rules matrix, 357

S

Saggital diagram, 10
Sensor, 357
Sets
    classical, 10
    fuzzy, 10
    theory of, 10
Simulation, fuzzy traffic control, 262
Software, fuzzy logic, 112, 374

T

Tactical aviation navigation, 363
Tautologies, 36
Thermal systems, fuzzy control of, 112
TIL-Shell fuzzy software, 112
Togai fuzzy software, 112
    Fuzzy-C expert systems, 112
    fuzzy development systems, 112
    single-board fuzzy controller, 112
Traffic, fuzzy control of, 262
Tuning, in fuzzy control, 374

U

V

Viscous, 357

W

X

Y

Z

AUTHOR INDEX

A
Akbarzadeh, M. T., 232
Alang Rashid, N. K., 374

B
Baird, R., 392
Baker, C., 363
Barak, D., 329
Bisset, K. R., 262

C

D
Duerre, K. H., 181

E

F
Fotouhie, H., 171

G

H
Heger, A. S., 374

I

J
Jamshidi, M., 1, 112

K
Kelsey, R. L., 262
Knight, R. J., 232
Kristjánsson, E., 329
Kumbla, K. K., 292

L
Lashway, C., 279

M
Marchbanks, R., 149
McCullough, M. K., 357
Miller, D., 279
Moya, J., 292

N

O

P
Parkinson, W. J., 181
Peterson, S., 279
Plummer, K., 329

Q

R
Ross, T. J., 10, 36

S
Sayka, P., 217

T

U

V
Vadiee, N., 51, 86

W

X

Y

Z
Zadeh, L. A., i

Please send me information on:

_____ FULDEK© Software for Fuzzy Logic (full version)
_____ FULDEK© Software for Fuzzy Logic (student version)
_____ FUZZCtrl© A Fuzzy Control Toolbox for MATLAB™

Name: _____
Institution: _____
Address: _____
_____

Telephone _____ Fax _____

TSI Enterprises Inc.
PO Box 14155
Alberquerque, NM 87191-4155
USA
Telephone: (505) 298-5817, Fax: (505) 291-0013

Please send me information on:

_____ FULDEK© Software for Fuzzy Logic (full version)
_____ FULDEK© Software for Fuzzy Logic (student version)
_____ FUZZCtrl© A Fuzzy Control Toolbox for MATLAB™

Name: _____
Institution: _____
Address: _____
_____

Telephone _____ Fax _____

TSI Enterprises Inc.
PO Box 14155
Alberquerque, NM 87191-4155
USA
Telephone: (505) 298-5817, Fax: (505) 291-0013

		Please place stamp here.

TSI Enterprises, Inc.

P. O. Box 14155
Albuquerque, New Mexico 87191–4155
U.S.A.

		Please place stamp here.

TSI Enterprises, Inc.

P. O. Box 14155
Albuquerque, New Mexico 87191–4155
U.S.A.